MOVING WITH THE ELEMENTS

An Ice Cube Press Publication
Of Solvent Fiction and
Restoration Nonfiction.

Moving With The Elements

Steve Semken

The Ice Cube Press
North Liberty, Iowa

Moving With The Elements
fire, water, earth and air.

First Edition
2 4 6 8 9 7 5 3 1

The Ice Cube Press
205 North Front Street
North Liberty, Iowa 52317-9302
E-mail: icecube@soli.inav.net
Web Page: Http://soli.inav.net/~icecube
Orders and comments welcome.

Library Of Congress Cataloging Number
97-93453

ISBN 1-888160-35-7

Manufactured in the United States of America

Portions of this book appeared in slightly different forms:
"Plastic Puddle Drops," previously appeared in *Magic Realism*
"Moving With The Elements," appeared in *The Source Magazine*
"Bright Dynamite Creek," appeared in *Potpourri*
"Yolk White Sky," appeared in *Trapeze Magazine*

In honor of creation
and earthly elements
flying, forth
and forming,
through weather,
wind
and water.

Many shouts of "thank you" need to echo out in all directions to those who helped in one way or another. As always any faults are entirely of my doing and not theirs:

Laura, Kit, Wendell, Andy Driscoll, Emerson Blake, Joni Palmer, David Rothenberg, Gene Logsdon, Denise Low, Paul Steinbrecher, Matt Welter, Jack Ozegovic, Siobhan Wolf, Ethan Hirsh, Walt McLaughlin, Daniel Bergeson, James Koller and Mr. Joseph Tucker, a true warehouse of story.

Moving With The Elements

Moving With The Elements

Cirrus clouds are wrapped lightly around my neck. I am standing on a peak of the Great Plains, looking out across long drawn out buttes toward the north. There is a delicate light now as the sun rises. The clouds look like ocean waves crashing into the soul of a rainbow. Peering deep in the middle of these brilliant morning colors I find myself a prospector, searching the landscape for elements, panning the air for gold. While searching, and looking, and wondering I become intrigued by such vast and creative variety. There is the rapid movement of the heart of a rabbit, the crackling ruffle of cottonwood leaves, the glide and flick of a garter snake, the twisting forms of walnut branches in search of beams of sunlight. There is the leap of a frog toward water and the changing length of shadows while the day goes by. Watching, breathing, thinking, the clouds around my neck slowly disappear and then there is the slowing flight of a small kestrel which lands nearby. I look at the bird for a message, for some reason why it has landed this close to me out here on this peak of the prairie. What comes to me is a set of eyes, something as intense as the invisible, hot, center vapor of fire. The kestrel's glare drills me deep, like a stone agate, or the surface of still water.

While moving down off the peak, a strong wind moves through, stirring up a thick layer of dew which has formed on the blades of grass over night. This gust of dew-filled air moves across my skin and a slight chill comes over me. I lift a jug of water to my mouth and it is an easy thought that comes to me as I drink. I understand that I have indeed struck gold. What becomes clear is how all of nature's elements are shining bright and that creation is constantly in full swing.

It's no mistake the "Good Book" begins with creation stories. For all the elements, and all of the world, and all of thought, and all of everything existent is formed through the motion and mixing of the four elements. A sudden, big bang seems unlikely in many ways. A large blast of the sort that makes a world would have to be steeped, leaked, creeped, dragged, swirled and slowly, like the silent stalking of a blue heron, formed into an eruption. Quickness and high speed seem opposed to creation.

So there is a red sunset in my veins. So I am becoming a prairie alchemist. So the earth is rooted in loamy, deep, dark and supporting earth out here in the heart of the tall grass prairie. The kestrel that is near glares at me again. Then lifts off, heavy beats of wings away. I walk slowly to where the kestrel has been, expecting to see something important. Hoping to find some clue that may better explain the world, but there is nothing. I look out to watch the bird's flight, but it has twisted into the morning light and blended invisibly with the sky. All that's before me now is the wide expanse of the prairie. Tall grass tipped golden with sunlight, swaying together with the wind.

Sky Water

Sky water arrives clear and clean like ice. The morning starts out with a deep, full colored blue sky and the land has been covered with a cool and drifting layer of fog. Birds are out flying in tight and noisy groups. A northerly breeze is moving crisply through the air and thoughts of early spring come to mind: plump, green pea pods, bright red radishes, sprouts of asparagus hatching from the soil. Today, sky water clouds are spaced out evenly into the distance–soft puffs of white, moving briskly overhead. The day has begun as a surrealist painting, the clouds overhead seem like giant ice cubes.

It is very quiet outside, each cloud seems to need a space of its own. I continue to stare up in the air. It is odd, a sky so cold and clear in the middle of this summer which has been so hot and hazy and muggy. But now there is a cold front in the air and I pick out a single cloud and watch it's shape and movement. It glides so slowly I am convinced that the cloud is patience in disguise.

While watching the sky I slip into the past and envision that I am raising the lid off an old deep freeze. I am out in an old weathered barn surrounded by aged lumber and cattle equipment, and there is cold steam drifting out as I raise the lid. There is a chorus of crackling in the ice box when the outside air invades. It is a silent world in the deep freeze just as it is in the sky today.

The freezer I am remembering was my grandfather's. Many years ago when I used to spend summers down in Mexico I would join him and we would slowly walk across the dry and dusty soil, out to the barn and open his freezer. He kept the freezer locked and whatever he took out was always good: fish, fishing bait, his sausage specials mixed with hot peppers, cheese and eggs. The freezer held sky water, held sacred items, it was a sacred space. Never was the freezer mistreated, never was it made fun of. It was always left alone and was trusted. Today, up above me, the silent, blue, chilly clouds in the sky become goosebumps. I think I hear my grandfather call me out to the freezer barn. The hair on the back of my neck rises, tickles, bristles.

Looking up to the sky water I decide that maybe it is possible, maybe I could be ten years old again. Maybe I am walking out to the old barn to help my grandfather take sacred items out of his ice box. Perhaps sky water is running through my veins.

I Will Drink Clouds

Walking along a spring fed stream in northern Iowa, I pause and look into the water and see clouds resting lightly on top of the moving bed of water. I touch and glide my hands through these clouds, but they don't break up, don't float away. Instead they only ripple and form into my skin right along the water's surface.

There is space, just above and just below the water where reflection is equal to reality. It is this space that captures me. I dip in a cup to scoop out a serving of clear, fresh, liquid. I look into this small portion of water and see that the clouds are still there. Tipping the cup to my mouth I drink, deeply, and can feel the water chill my stomach, literally getting a cold gut.

Looking back up into the bright blue sky I start to fly. My arms and legs feel no more than flecks of vapor. I feel light headed. After all, there are clouds inside me now.

The stream appears again. I imagine that I'll be far away though, somewhere high in the sky, looking down at the stream, but instead I am looking up out of the water. Then I hear a trout jump. I am confused. Something has happened, I am short of breath. I get the feeling that I have just fallen from the sky.

I look back into my glass of water, covered with clouds, and drink again. Suddenly, it's not so hard to learn about evaporation and know how to fly. To experience what it's like to be precipitation and fall from the sky. To be a still and silent reflection atop a northern pond. To be squeezed and condensed from water into ice. To rush with adventure down a raging river. Today, as I look at this spring fed stream and drink, I understand water in all its forms.

Cloud Recipe

There is a thin light over my head. A wispy chalky blue color up between three layers of clouds. The highest clouds in the sky will be the weather in about two days, the middle clouds tomorrow's weather and the low clouds are here right now. The air is in motion and the color of the blue sky keeps changing shades. I spot the first leaking of rain in the distance and the wind has just picked up and is blowing in my face. I notice that the birds have stopped singing and, in fact, only occasionally do I see a bird and if so, it is fluttering, out of control in the heavy, humid air.

There are no animals out or about. There is only stillness. Yet, as if to test this stillness, there is a single dog barking hollowly at the end of a chain.

There is a cold front, a falling mass of chilly air gushing in. A low pressure system, now so low it could hardly be above anything. At some point, as the day becomes a deeper and deeper gray, the dog stops barking. He sits down and sighs without regard to dignity, ruffling his lips, spitting and sighing loudly.

I look out into the eastern sky and see one last narrow glint of slate blue. Even as I look at this, even as I think of this mutely, off colored blue, it is gone, taken abruptly away by the wind.

Right then the idea comes to me, that where clouds begin there must be a large body of water. I cock my head a little with this thought. I believe I can hear a hummingbird nearby, or perhaps my mind is just racing. Then the sky splits brightly for a millisecond. Then the wind picks up. Then there is a crash of thunder and shortly after this, water begins to fall.

Standing in the rain, I wonder if where I am now is where clouds begin again.

Oak Knot Bridge

The oak tree is a bold carrier of the earth's elements. It stands straight, condensed in compact grain. Not too tall, not too short. The oak tree inhabits both soil and sky. It is supported by the weather and assisted in the formation of its growth by the wind. Although not a friendly companion with fire, other creatures that enjoy firewood and heat would disagree. In fact, would probably say that the oak is fully realized through fire.

The oak tree is going with as much speed as a hardwood tree can toward the sun, branches moving toward the golden fire in the sky. At the same time, the roots of the oak tree are digging down toward the burning and prehistoric lava in the center of the earth. The oak tree wants to get to fire. It's trying in all directions to bind water with fire.

It has been said that each large boulder contains a forest, each small stone a tree and each grain of sand a leaf that has fallen. Looking around in the forest, it becomes possible to believe that the purpose of the oak tree may be to join the sky and the earth together. Let it be known that when the center of the earth meets the surface of the sun, a knot is formed in the trunk of the oak tree.

Childhood Weather Addict

When you notice a little child standing along the shore of the ocean, looking at the sunset, looking at the flight of sea gulls, watching the height of the waves, maybe noticing how strong the wind is, you are not watching a child daydreaming, but a child communicating with the natural world. A child learning to understand the elements on their own, not relying on the help of grown ups, or older kids for explanation. This is where the marrow of spirituality can begin. Being able to follow the variety of colors in the sky, watching the clouds float in free forms, feeling and observing the breaking of waves, all reveal the spontaneity of nature. Observing the unexplainable is to catch a glimpse into the cathedral of the sky.

When I was little a mighty soothsayer floated down on a cloud, joined me in my backyard and charmed me with the desire to be a weatherman. I was riddled and tempted with the mysteries of wind, rain, rainbows and the sun.

In Iowa I had heard about weather radios, and wanted to get one, but it wouldn't have done any good since there were no weather stations at the time. My lack of a weather radio provided a need to learn the weather directly. It made me have to observe the natural world without help. I would stand at the edge of the woods and watch the birds fly, look at the colors of the sunset, feel the wind blow and consider the formation of clouds. I wondered why the world was constantly changing. Wondered what was making the weather. Was is it a person, a wind, a God?

One summer when my family went to Sullivan's Island in South Carolina, along the Atlantic Coast, my Uncle had a hand held weather radio and I borrowed it and listened. Quickly the radio became uninteresting though, it seemed like it was playing the same thing over and over again.

I hated to admit it, but after only thirty minutes, my wish for a weather radio was dashed. It seemed useless and boring. I decided then and there that I liked the guess work of weather more than the reports on the radio. Weather from a radio just didn't seem natural, or magical enough.

Being able to understand the patterns of the sky without the help of a radio, or television is an admirable and intimate quality, as well as a skill that can never be mastered. To perceive that Midwestern weather patterns in Iowa usually move in roughly four day cycles just takes time to realize. I have barely begun to figure out what the winds mean, what most cloud formations indicate, but over time, patterns and predictions and understanding does increase. There are times when a person can smell rain in the air, when one can observe a change in air pressure through the flight of birds.

Weather controls our moods, our language, our locale, our boredom, our frustrations and also works as a good excuse for erratic behavior. It is no surprise that many people commit suicide, or harsh crimes in the dead of winter when cloudy days are strung on and on together.

The weather is real at each moment, it is a river above our heads. The natural and spontaneous creation of weather, the drift of clouds, the development of mossy, soft and vibrant soil is in conjunction with mist, rain and shadows that fall from the sky. The world is the truth as told by warm fronts and cold fronts, of high pressures and low pressures. The weather is a good book to read. The weather is a cathedral of smooth and varied light.

Yolk White Sky

It hardly seemed necessary to continue mentioning that the sky was falling anymore. Perce and myself had been collecting chunks of blue sky in our tree house for over a year now. The only reason we still had any doubt about the sky falling was because we had both asked our Dad's if it was true, "Was the sky really falling?"

"No! Now go away," they both answered.

Still, it was obvious that blue stuff was gathering on the ground and at night it was easy to see that several constellations were missing from the sky. Perce also told me that, although he had not seen what it was, he had heard a loud screech last night. When I asked him what he was getting at, he said that when things fell from the sky real far they made a loud screaming noise. I understood what he was getting at and nodded.

The next day, on the way to school, Perce shrugged his shoulders and said that he guessed the tree house was about as full of sky as it would ever get. I looked in to see what he meant and found that he was right. I also noticed that a few clouds had been trapped between some of the pieces of sky, and in those spots there were tiny rainstorms, downpours no bigger than a dog's paw, taking place. Perce walked over to the clouds and with a grin on his face, picked up a tornado, about the size of his fist, and began dancing in circles around the room with the tornado clutched closely to his chin. There were also stars and even constellations hovering close to the ceiling of the treehouse. A few planets were orbiting the room, as well as thunderheads spewing forth lightning bolts, making our treehouse flicker bright as a lightning bug.

Immediately, I was reminded of my Uncle who had died a year ago. He had danced ballet wearing a black suit, in a theater hall, on a polished, white, marble floor, underneath a yellow light bulb. I had been told he moved as if he were walking on air. He was fluid, graceful and gravity never seemed to get in his way. He would go around and around, toe tips to fingers, in twirls, alongside music made for heaven. My Mom called him Cloud, because she had thought of him as an elegant cloud drifting through a vivid, blue sky. I'd always wondered what had happened to Uncle Cloud and now, standing in the tree house, I knew.

People always tell me that I have a habit of drifting, of wandering, of wishing, but nothing like Cloud did. It seemed obvious now, I guess, that Cloud had died and moved up into the sky, up in space beside the milky way and was busy clearing out a white dance floor around the sun. I was certain that after the sky had completely lost it's color and fallen apart that Cloud was going to begin dancing on what would be a clear white roof above the world.

After I got done thinking about Cloud, I noticed that Perce and I were floating out the door of the tree house on a shooting star. We were zooming beyond the horizon. It seemed we weren't alone either. After we had come to a stop, we heard the chattering and whispering of an audience gathering. There was a group of people in anticipation of a grand performance. Perce was tapping my shoulder, he was pointing to a stage, where, just as I thought it would be, my Uncle Cloud was beginning a dance, out on top of the yolk white sky.

Under The Lights and The Open Air

If anyone can tell a sports story it ought to be me. I earned nine letters in athletic competition in junior high, getting the rarely awarded C-Hawk award. Earned six more in high school. Have played nearly an infinite amount of basketball, softball and baseball. I have been on rowing teams, frequently fly fish, play frisbee, hacky sack, boating, hiking, skateboarding, running, skiing, biking and find myself reading over the sport section in newspapers unconsciously. Yet, I am miraculously sports-less when it comes to telling a sports story. I know that when I was very little I used to hold imaginary world series games in my backyard, tossing a tennis ball against the back of our garage, putting dents and holes in the siding. I was like most kids during summer vacation, pretending to be ten players at once, in game seven of the world series. Or I would act out the All-Star game, pretending I was Luis

Tiant, Tom Seaver, Vida Blue, Catfish Hunter, Don Sutton, or Rollie Fingers. Not coming from a strong sports family, my early sport stories were mostly formed in my head.

Sports in the outdoors. This is what interests me now and in the open air, only one sport comes to mind, and that's the game of baseball. Pull this image into your vision. It's the Northern League, out in North Dakota somewhere, let's say it's the Big Skies vs. The Horizons. The national anthem has just finished and there is a minor league ball player standing on the field, with a bat over his shoulder, looking out into an evening sky. Waiting. The stadium is bright, so bright, a crystal of white light and beneath all the light is the green, ever-so-green field. The players are all dressed in clean, white uniforms, and the field and the realm of space illuminated under the light is comparable to a vast and holy church. When the players are warming up they talk with the officials, they toss the ball around easily, so accurately. The pitcher is popping the ball into the catcher's mitt with the precision of a surgeon. Suddenly, the yell of "Play Ball!" goes out and the stadium hushes. The ritual is about to begin. People settle down, eager to observe.

The sky is clear, there are stars above. Beneath God and the Holy Ghost the game is under way and the fourth hitter walks to the plate. There are two runners on base. The first pitch is thrown, "Strike!" The official likes to yell, everyone knows what's going on with him. The batter stands back and looks around the field. One strike is nothing. He looks out to the fence. He can hear yelling in the background, but it is vague, no more than a blurry noise. He thinks about nothing in particular. He stares at the fence again, steps back in the batter's box, takes a light swing, sets himself, then looks at the pitcher. Ready. Then everything slows down. The only thing standing between the batter and a homerun is sky. Pure, open, air. The pitcher moves his front leg, lifts his arms, rocks back, then rolls forward and lets loose a fastball. The batter steps forward, twists his hips, moves his wrists and swings his arms through the ball. There is a ripping snap, the sound of wood popping and as everyone's eyes focus, the ball is flung into the air. Then there's the roar of the fans as the ball is lifted up higher and higher into the hemisphere. Everyone looks into the sky with hope, and as the ball hangs in the air, there is a brief moment of silence, right before it becomes obvious that not gravity, not wind, not the force of the full moon, not the spinning of the earth, or the big dipper in the night sky, nothing, nobody, or anything is going to stop this ball from continuing to fly over the center field fence. The crowd is ecstatic and yelling. They are outdoors, they can yell all they want. Kids slap hands with their dad's as if they were best friends, there is a brief moment of magic in the air, drinks spill in the aisles, but no one cares because they have been carried across the universe and through outer space by the flight of the baseball. The night sky is alive, it's a HOMERUN for the home team out under the bright white lights where the greenest, green grass grows underneath the big, wide, open, night light sky.

Morels

People like to talk about mushrooms, the family fungi is a mysterious topic. Where to find them is akin to surprise. A giant puffball rests easy and silent, yet takes a looker with a deep gasp. Mushrooms are a personal thing and most of the talk about them is not very revealing. If you are listening closely, repeating back what you're being told, you'll figure out that morel mushrooms, in particular, are a secret. Like a good fishing hole, no one will really tell you the intimate details of their location.

One time, someone mentioned to me that they were thinking of training their dog to smell morels. We both kind of laughed. I went one step further, took this guy by surprise. Grinned and said, "You could be like me. I'm working on sniffing them out myself. Smell kind of like sweet black dirt and leaf mold." The guy nodded back, and said, "Yeah, well that could work too. I guess."

Now, although people say that mushrooms are spores, I would disagree. I think of them as the birth of rocks. Soft shell stones. This comes to my mind since new stones are heaved up out of the earth with each new spring and rest on the land like miniature totems, like monuments to the way the native land was meant to be. Heaved stones are the earth speaking up, giving up secrets to the sky. In much the same way, morels rise to the surface of the earth as if in tribute to the return of soft rain, loose soil and cool air.

Morels exceed almost all labels. They can be found just about anywhere. Lots of people tell me, "I looked all over in perfect, mossy, mushroomy places and never found any." I usually respond, "Try someplace where you're almost positive they wouldn't be. I've found them in the middle of a sunny field before, you just never know. Most of all, start looking earlier than everyone else, keep looking longer than everyone else."

Morels come out of the ground as it heats up after the last frost of the year and when there have been a few good, misty, long, rain showers. I tend to search the regions where I can find sycamore tree leaves melting into the earth, and if this occurs on an east sloping hill, or in a small draw, then usually I am in luck. However, I'll admit it, I never pass by a place where there's bark laying at the base of a dead tree, as such textbook spots do hold a morel every now and then.

One more thing, there is a little rule to follow if you are hoping to find any morels. If you aren't willing to stand and walk in the rain, not willing to kneel down in the mud, then you won't have what it takes to find any morels, and for that matter, you don't deserve to find them. It is absolutely necessary, if you're going to enter the world of the morel mushroom that you sink into the soft, wet earth and feel the roots of spring growing. This makes you part of the taste, part of the smell, part of the mud and part of the clouds that create them.

If you wanted to read about finding morels, wanted to really know about these surprise attacks of spring, you would need a strange guide for help. The book would have to be quirky, it would need to have an envelope in it with dirt samples and various test tubes containing odors. You would need to have a translation guide to cope with the amount of hearsay you will encounter from others. The book would have to be entitled, The Book of Morel: A Hunting and Hearsay Guide (Talkingman Press, 2001).

Short, Quick and A Lifetime

I'll call him Henry. He had a toothpick stuck in his mouth and ended up speaking small, quick, true words to me. I had heard of him from others, that he was a man of action, not talk. I had been informed that he hadn't spoken more than four words at a time since 1956 when his lips and lungs had been burned up while lighting a cigarette, with a blow torch at work. Seems his co-workers wanted Henry out of the business and left the torch setting on full bore. Since then, Henry just didn't need spoken communication, he got along just fine with a smile, a frown, a sneer, moving his head and lifting, or lowering his shoulders. With some he had never spoken even one word, he didn't believe in wasting his time. Some thought he had been picked-on by people too much. Folks would ask him if he had spoken yet and he'd say, "For what?" real quiet like. And they would laugh, "gotcha."

Well, Henry pulled me aside one early spring evening, when the town was nearly vacant, and opened up for a good three minutes, telling me something I hadn't expected, which is the way I've found many old people are prone to doing. He told me about stones and family. He said that his family was immovable. That they were held together like a rock. Nothing could separate any of them from one another. When any one of them would die they made a point of laying such a heavy tombstone above the body that the only thing that could escape was the soul, not the body, or the memories. He said that being in the dirt, held down with stone, was how everything else stayed behind so that the passing family member could remain nearby, on the family farm. Rocks ensured that future generations could live alongside past generations, in a real and growing, mutual way.

"The soul we can do without. The memories, we need those to stay behind. The memories we hold on to and can use," Henry told me.

The whole time he spoke, I had a feeling that he wasn't really talking to me, but to someone's memory in the earth and that I just happened to be nearby. I also noticed that all of his speaking had caused the edges of his lips to start bleeding. I nodded toward him in appreciation of his words, honored to have been there. Walking away I felt a new connection with language.

I found that I didn't speak for nearly two days afterwards. Instead, I picked up smooth stones and held them in my hands. I thought about all the people I had known over my life, thought about those who had passed away and wondered whether all their memories were still available to me. I found that this sort of thinking kept me very still, very quiet, very connected with life.

Early Season, Arctic Shotgun

It is cold out. It is January, and I have been skiing the Ice Age trail in Wisconsin. I am tired and leaning over my poles, looking out from a high ridge above Devil's Lake. There is a frigid cold front coming through, and I imagine that the winter night, as the sun goes down and the wind picks up, will be amber colored just like brandy. It will be a silent night. As the arctic front clamps down, the whipping wind so dry, raising and dropping dry snow, churning the flakes into a crunchy substance, into snow so puffy, so tense, so tight that it creaks under foot.

I look deep into the forest and spot a bird in the carved hollow of a tree. It is sitting motionless and hoping for warmth, for an end to the cold wind. It is hard to understand this bird's wildness, the stoic strength that it must need to survive, such solitude and isolation. I can merely look at the animal, so alone. I want to pity it, but know better. It is so much a creature of the wilderness that I can't possibly relate. I respectfully move on by.

It turns out that it is deer hunting season this cold and frozen day. I feel like the hunted in this cold weather. Feel as if there has been a scope on me the way the cold is attacking my lungs, my feet, my fingers. The cold is trying to suffocate me, but I survive. I ski to the end of the trail and am able to return to town. Driving slowly across crunchy snow, past silent homes with smoke coming out of their fireplaces, I relax. I have mixed emotions about going indoors, but also realize I have no choice. I can't survive in this weather. I understand that the winter humbles both me and my community in the north. It is valuable to respect the wild and I do so as I leave the cold, shotgun barrel wind and stand beside a fire indoors.

Sky Printing And Land Mirroring

The sky is the imprint of the land and the land is the imprint of the sky and the space between the clouds is as much sky as it is land. Perhaps everything is just space.

It is pleasing to think that the sky dips down and lays an impression on the earth and there you have a great prairie, full of wind swept grassland, mile after mile of slow angular buttes rising and falling into the distance. It is just as comforting to know that the land can rise, reach to the sky and form clouds–long, dark lined fronts pouring rain in patterns that resemble stalks of blowing grass and seed. Never mind that this is all completely beyond comprehending. Think a mantra instead: sky water, shadow dirt, sky water, shadow dirt. Rest on your back and look. Take ten minutes to try and decide if the land makes the sky, or if the sky makes the land. Or maybe ask, how blue is blue, or how solid is a cloud?

As I look up and across the vastness of South Dakota, I see the sky as never before. I mean really, it's totally outlandish this much space and air and color and cloud variety within eye sight. I'm sure each spot of sky possesses individual differences, but I am only able to see it all as a whole. It is so much larger than my meager, practical, human imagination. I think of wide and big, my best words for expression, or I speak as the educated–vast, expansive, titanic, heroic, gigantic, but all are just synonyms for wide and big. I know you should never say never, should stay away from forever, but I decide that I can speak like this regarding the sky. I realize that if the galaxy is growing at the rate of light then, sure, there's probably a moment when forever isn't true, but not as far as I can prove.

I am in deep space now, in Dakota. Alongside the surface of a smooth chalky creek, in the badlands. The sky is full of possibilities and is twisting my mind into dreams. How many of our nation's astronauts have come from the prairie states because they felt at peace with the sky, or because they just had to know the air, space, forever, better? It makes sense to want to get up in the air and believe in the vast and open world of the prairie. To prove that walking across the Great Plains on a clear, star filled night is really no different that flying through outer space.

Although I have been looking, it feels like I am just now beginning to open my eyes. I am determined to go back to the easy ways of thinking, to the way I used to believe: that the land was different than the sky, to when I thought I walked on dirt and couldn't fly, when I thought water was either a lake or a cloud. I am determined that I will be like this again, but when I look at the world out here on the prairie, the land and the sky blur together. I can't tell them apart. I know it's not possible, but I do see a green sunset, I do see orange grass. I do see red trees. I can see a purple lake. I take a step forward and I think I touch the sky. I tell myself that the grass in my hands is wet with ground water and not from clouds touching the earth. I tell myself that I am not hovering above the ground, but swimming in a lake. I am not completely convinced though. I look down at my shoes and they are covered in mud. I realize that I have no idea where I've come from, or where I may be going.

Bright Dynamite Creek

Things could have gone opposite of the way they did when Tiffany and Tony Spencer pulled out of their driveway. They were going to Diversion, North Dakota with their parents for the summer. As they drove away things weren't so good in Madden, Iowa. Bobby Swartzendruber was sad, and I was already bored.

There were still people left in Madden, don't get me wrong, but when the Spencer's left town, as far as me and Bobby were concerned there wasn't a soul left. Bobby and Tiffany were both seventeen and going out together. Me and Tony were both twelve and best friends. When the Spencer's left, what they had looking back at them were me and Bobby standing on their front lawn, waving good-bye real slow.

Bobby looked at the car disappearing and started talking, "Dang it, finally get done with school, but turns out vacation is even worse than school could ever be." Bobby looked down at me, "What you going to do now?"

I was surprised Bobby was even talking to me, and then I was even more surprised when I actually started telling him what I planned on doing that summer, "I'm going to play with made-up friends, I guess. Going to call myself Levi Toughskin, Man of Untearable Jeans."

Bobby sort of looked at me strange at first, then he laughed, in a grown-up kind of way, heavy with his shoulders like he thought I was stupid but cool at the same time. Then he said, "Well, Levi, you can call me Rico Rembrandt, Caretaker of the Color Blue."

As soon as Bobby said this I knew the summer wasn't going to be as bad as I thought it might be. I looked back up at Bobby, we caught eyes and then he walked off to his house. If all else, even if I didn't see Bobby again that summer, at least I had a new character to play with.

One day during the summer, down by the creek, while I was playing that I was the strong and mighty Levi Toughskin, I discovered something. I discovered that there was a place under the water where things changed shape. I found this out by dropping a firecracker down the middle of a hollow whirlpool in the creek. I dropped the firecracker down the middle of the funnel and then after a little while I heard a big, loud, hollow bang and saw some water spew up a little down stream. Then I floated a feather down the center of the whirlpool, and after about thirty seconds a big, bright Blue Jay came flying out of the water about ten yards away. This is when I decided I needed to talk. So I ran to Bobby's house and told him. He didn't believe me at first, but I led him to the creek by tugging his arm and begging his mind. When I got him to the creek I did the same two things, the firecracker and the feather. Well, Bobby believed me then.

Being older than me and wanting to prove it, Bobby had a serious forehead about what we should do. I told him we should just keep dropping stuff down the funnel and see what happened. I could tell Bobby thought this was too easy, but he agreed anyway. He started. He dropped a pumpkin seed down the funnel and after awhile a pumpkin popped out and floated down the creek. I threw a handful of pumpkin seeds in and next thing we knew the creek was covered with floating pumpkins. Then I got another idea. I scratched my name on a chunk of rock and dropped it down the funnel. That didn't turn out to be so good. I fell in with the rock about two seconds later. Pretty soon I was swimming in the creek. "What happened?" Bobby asked. I told him I didn't know, just seemed like a bunch of bright lights.

Bobby did the same thing. Dropped his name down the funnel and I watched him disappear and then he got fired back out of the water, floating down the creek. "What happened?" I asked him. He told me he didn't know either, maybe he was swimming with a school of fish or something.

We both got big grins on our faces and probably dropped our names down the funnel fifty more times each. Whatever was going on it was fun. We would pop out of the water and be laughing so hard our cheeks were sore. I don't think either of us understood why we were so happy, maybe just the surprise of the water was addictive, some sort of victory for us. As the sun went down, I told Bobby I was going to go home and write down a lot of stuff on rocks and then come back tomorrow. "Yeah, me too," said Bobby. Then he said, "See you tomorrow Levi Toughskin."

"O.K. Rico Rembrandt," I said back. I caught Bobby smiling when he turned around to walk home.

When I got home I wrote what seemed like six million things on rocks. I wrote money, fun, water, Tony Spencer and Levi Toughskin to name a few. The main thing I wanted to drop down the funnel was Levi Toughskin. I wanted to see who my made up character looked like.

Bobby was doing the same thing. He was writing stuff on rocks and paper. He wrote less things than I did. Mainly he wrote Tiffany Spencer. He wrote her name sixty-two times. He also wrote Buick and Rico Rembrandt. He told me so.

During the night I didn't sleep. The thunder kept me awake. I also got scared because the wind was blowing so hard. One of the silver maple trees in our yard even fell over. Most of all I was thinking about how high the creek would be running with all the rain that was falling. I was thinking there could be problems. Bobby was thinking the same thing. He told me this, too.

I was at the creek as soon as the horizon began to show. I was right. It was not good. The funnel from yesterday had been washed away. I started throwing my rocks with words on them into the creek. I was suddenly embarrassed about what I had on them: too many wishes. While I was tossing the rocks in the water Bobby showed up. He was mop headed and sleepy. He looked at me, then at the creek, then back at me to figure out the news. I just stared at him. There was no way to hide the truth. "I thought so," he said. He sat down heavy. I thought he might cry, but he didn't.

"What did you write?" I asked Bobby. He told me.

"What did you write?" Bobby asked me. I told him.

We thought it was funny that we had both written down our imaginary names. I guess we were both hoping to find out who we really were.

Even though the day had turned out bad, I played I was Levi Toughskin for the rest of the summer. I think Bobby continued to be Rico Rembrandt as well. The reason I think this was because we both kept seeing each other, off and on again, along the creek that summer. It was obvious that we were both hoping to find the same sort of whirlpool in the water again. After all, a person needs good ideas to get through bad times.

Amidst A Tomato Sunset

A mouthful of gravel dust enters me as I cruise down the country road that goes by my front door, on past McPherson's Bridge, past the Hawk migration point, past Green Acres airport and on into the wilderness refuge. There I will rest with the sky, on top of a large boulder that has been absorbing the sun's heat all day. From here I will walk to an abandoned barn and watch shadows lengthen and roam the surrounding area. Shadows which search all over the ground with their eternal darkness. Chilly shadows lengthen and are cast across corn. Some ends of the corn are beginning to turn brown, fall weather is coming.

Yesterday I was accused of setting the air on fire because my tall, wide, fully alive tomato plants have reached up and touched the sky. They are afraid of nothing. They are mighty, they are all powerful and wish to join with the sun. The plants want to place a large, brilliant, red Big Boy tomato up in the hemisphere, want to experiment with the concept of orbit.

I hear there may be a light frost, a record low for a September day on its way. I am slightly in mourning as I foresee the end of the tomato plants which have become such a part of my life these last six months. I wonder if the sky isn't going to be empty when they are gone. Yet, at the same time, I am anxious to get started on the other half of the growing cycle, anxious to get the compost pile into high gear from the remains of the summer's growth. I take great pride in my soil. I am determined to turn the naked, poorly managed city soil I inherited into lofty, loamy, healthy, natural earth. I want to smell the faintest odor of rhubarb when I hold handfuls of this soil to my nose.

As I think of this past growing season, of my many peas, beans, canteloupes, corn, basil, beets, tomatoes and cucumbers (a cucumber patch now made infamous with the label of Big Bertha by guests who were amazed to find that we had canned nearly thirty-five quarts of pickles in a four week period off two vines), I am once again amazed at the interaction between soil, water, air and light. Thinking back, it seems a long time ago that I put these plants in the ground, not just within the last few months. Every year the development and growth of my garden plays an important role in my life. Each garden, each growing season, takes on a character of its own. Some years the green peppers are miracles, some years they are curiously sluggish, sometimes they are nonexistent.

The past few years I have added a new twist to the growth cycle. I have taken to creating fruit wines. Cutting rhubarb, picking blackberries, squashing wild cherries, plugging corks into bottles. It is an obsession of sorts, but not for what others seem to think. It is not done in order to drink, so much as it is to be part of the growing season. The wines are a celebration of nature, an intimacy with creation. However, I should probably admit that the flavors of rhubarb and blackberry wine are monumental. I can assure you that this aspect is not discounted.

Being so involved with growing, with building, with making wines, I am curious when I am asked if I would have preferred to have been a scholar, if I am happy even though I am not a "professional." I am questioned about not having a career with a practical, financial, occupation. It appears that using my time to create wine and beer, to spend my time writing, designing and searching for trout is bad, or perhaps just too simple. That I have not committed to a larger cause than that of myself boggles other's minds. I suppose I don't believe a person is who they work for, what their job title reads, but how well they imagine and, more importantly, what they actually accomplish of what they hope and dream for.

I go back to my tomatoes. I hold one up at sundown and compare the color of it with the red of the sunset. They are nearly the same. I hold a handful of sun in my hand and this seems good enough. It is not far from reality.

Kansas Cloud Accumulation

If you had asked me what boulder glass meant a few years ago I would have said, "Huh? A glass eye, maybe?" I may have thought it was a Zen koan of some sort, an impossible riddle, if it weren't for a kid on the Great Plains.

The most magnificent thunderhead I have ever seen was in Kansas. I was in the northwest corner of the state, in St. Francis, when I looked to the south and saw a huge and towering, absolutely hemispheric reaching cloud accumulation of titanic proportions. There was lightning stretching all around inside it, like the veins on the back side of an oak leaf. All around the landscape there was simply stillness. The rest of the sky to the north was clear blue, empty of clouds, making the scene even more outstanding. Out along the streets in St. Francis people were standing beside their cars and trucks watching. Folks were gathered outside their houses and looking to the south, just like me. Considering this was western Kansas and that people were stopping what they were doing just to look at a cloud is really saying something. These people live out in the wind, the air and open space all year long. Passing clouds are just a way of life, no more of a deal than silence and drinking. You can imagine then, how massive this towering cloud must have been, it really was an incredible, sinister sight. It loomed, with the passion of an assassin, out in the distance. Every once in a while you could hear it roar, could sense it trembling. It could have its way with the landscape if it wanted to, but as night fell, everyone was glad to see it moving east. It was a beautiful thing as long as it was going the opposite direction.

Before the sun set, I heard some little kid ask his mother, "Momma, is that what daddy calls Boulder Glass, that kind of cloud that's dark and tall, looks mean like a rock, but you can still see through?"

The kid's mom just said, "I don't know, honey, I don't know. Let's just look, okay?"

I thought this was an incredible description of a massive, dangerous cloud. A cloud formation that became like an essential element for a period of time. Thus, it was in Kansas when I fully understood the idea of there being see-through solids.

Speaking With Manny

The day begins so evenly, so level. A bright, low sun casts my shadow nearly fifteen miles out in front of me. It is the longest day of the year, in the flattest part of Illinois that I could find, and I have come prepared with binoculars to witness my image elongated into the next time zone. I am amazed for it is absolutely, positively, one hundred percent clear outside today.

"If you were on a bike it would take an hour to reach the end of me," I say to a sudden stranger. He is an older man, plain and slow in pace, who stalls between his breaths. He tells me that he has an irregular heartbeat, but that over the last few years he's probably gained a couple days of life since he figures his irregular beat gives his heart an occasional rest. He also says that he has to move very slow now. It also appears that his eyesight isn't quite clear. He wobbles just a little when moving forward. He is wise though, I can tell, because it doesn't startle him that I am dealing with shadows. The conversation continues as I tell him that I was born in Michigan, have lived since then in Iowa, Idaho, Kansas and Oregon briefly. I tell him that there are some other places I might consider moving to, but not for awhile yet. I consider speaking with him about the weather, but he beats me to it.

"Good day for sun," he says, smudging the ground with one of his toes.

I say, "A great day for shadows, the best day in fact." It turns out that his name is Manny Sanguillen. "The baseball player," I ask?

"Oh no, no not me, played a bit of ball here in my hometown, but I am not the ballplayer." Manny saunters when he speaks, that is to say, he seems to be thinking about three or four things at once and only manages to get the words out of his mouth slowly, making sure to tell only one of the four stories he could be telling. I wonder, and almost wish I could hear, what that would be like, to hear four stories going on at once. I have a strange feeling that I could easily follow along and am tempted to ask him to tell me everything on his mind just to see what it would be like, but then I think, what if he isn't doing this at all and just goes, hesitatingly, slow.

He is a believable person, honest. I am impressed. He acts out the position of other people when he talks about them, like a stage performance, he roleplays in remarkable fashion. Then, while he is acting out and describing different types of men talking, the sun behind us approaches the perfect angle for lenghty shadows. I look out and I see the absolutely longest shadow of myself that can be cast. I am nearly eighteen miles long.

I interrupt Manny as I look out in the distance and say, "Manny, I have more length than the flattest space out here. See me out there on that hilltop? If it weren't for the rise in the land I would be out there maybe twenty miles."

Manny sticks his hand in the air and makes a shadow bunny, it takes a couple minutes for us to find his shadow, but then we see that it is being cast on the wall of a barn about six miles away. He tweaks the rabbit's ears. We both chuckle. I move my head and before I know it the sun is rising and my shadow is coming back to me.

Manny drops his hand and shakes mine. For what I don't know, so I ask. Manny says he doesn't get to see people much in his life. He feels like he's no more than a piece of rock, sluggishly moving along, his walking stick in his left hand guiding him. It is a little strange he says. For a long time he was always running errands, doing chores, having to go into town to talk about business, to get equipment and supplies. Then, a few years ago, without a family around, without the necessary skills to drive, everything he got was delivered. What really got him wondering, about himself and human nature and pride, was why he always decided to go on a walk when he knew that his supplies were going to be delivered. He knew it was stupid to pretend that he was too busy to meet with the young lady that delivered him his groceries and his mail. He knew that he'd really like to spend some time with her, talking, but he also knew that he would feel vulnerable and embarrassed to have so much to tell her. So he avoided her.

He explains he wasn't expecting to see me standing here today. He says that his farm here, this incredibly flat spot where we are now, he calls echoes, Echoes Farm. He explains that the place has a soothing and repeating pattern to it if you give it a chance. He says all this twice and laughs lightly, "You get it, echoes, echoes."

I say that I am writing for the Ice Cube Press in Iowa, and he says that sounds cold. I say, "It's supposed to sound cool, but people tell me it's not professional enough." He scoffs at this and says, "Figures, people always have advice, not many do much of anything though."

"Or have individuality," I add.

He tells me how the worst thing he can think of becoming at this point in his life would be a statue. Not that he thinks there's any chance of someone making one of him, but the idea is horrible. He is afraid of life he admits. He never thought he would ever walk so slowly, rise so slow, breath so slow. He explains that his worst nightmare is that his arthritis is really an accumulation of gravel and that he is literally becoming a rock. "I don't know, a statue," he says, "that seems too unlike a person for me. I sure wouldn't want that."

I never thought of it like that before I confess. "How about becoming a memorial shadow?" I suggest, remembering what I was here for, but also thinking about his idea of not becoming a statue. I notice that I am speaking slower, trying to keep my different lines of thought from coming out mixed. Manny notices this and says, "You see, it's hard isn't it, this thinking and talking stuff. The mind moves quicker than the mouth, at least with me it always has." Manny says this isn't too deep a thought, but that he thinks the spirit of a man ought to be visualized, or if permanent, more like a shadow, twitching and changing with sun and weather. He says he likes my idea of a memorial shadow.

Manny suddenly gets quiet and looks pale, he has stopped breathing, then quickly he is back and he asks me what I am writing about. I tell him that I am studying the way weather works, that I am putting together a book on weather. I am hoping to write a modern handbook to weather and clouds and stones and masonry bits.

Manny pauses, looks out in the distance, then back at me, directly in the eyes for the first time, and I notice that he has a black patch over his right eye. I observe that the walking stick in his hand is made of bright yellow, hard, Osage Orange wood, and that it is trembling in his grip. He tells me that coming from a farming area he knows one thing, people are powerfully interested in the weather. It is most people's livelihood. He pauses, "Make sure you mention that somewhere," he tells me, "That weather is most people's livelihood, even if they don't know it." I nod my head to acknowledge this comment.

The weather, it seems to Manny, is always classic, and so it seems odd to him that I am writing a modern guide. "Why heck," he pauses, "has anyone even written a classic guide on the weather yet?" He starts in again, slowly, "I have always wondered if the weather isn't following a long repetitive pattern. Any chance you might try and cover an idea like that in the book too?"

I tell him I will put that in the book as well. Then I mention that the reason I am here now is because I want to figure out a connection between the longest day of the year, shadows and the weather.

Manny stops breathing again. I wait, and then, he's back and says, "That's good. When I think of people writing books and being serious, I never imagine anyone actually going outside and trying new things out, looking around and being alive. It seems like books are usually boring displays of words written to get folks famous and, from what I hear, a substitute for acts of teaching at those universities."

While he's talking the sun goes behind a long, lone cloud, as if to dare the sun on this, the longest day of the year. I realize that I am lucky to have had a perfectly clear, bright sunny day this morning. In the absence of direct sunlight I notice that Manny is gone. Not just kind of, not just mentally, I mean, really gone. I can't find him, and I am twinged with a combination of wonder, fear and embarrassment. I am wondering to myself if he really existed. Surely I am not speaking to myself. When the cloud has passed by the sun and it gets bright again and when Manny is still nowhere to be seen, I look to see if he is casting a shadow somewhere. Doing this I notice that even the barn which I'd seen him cast his shadow rabbit on just a little earlier is gone too. Even the hillside where I lost my shadow has vanished.

There is only one explanation, assuming that I am not crazy. On the longest day of the year, on this flat portion of the earth, there is more to see than any other day of the year. There is more pure matter, more life, more of everything being illuminated. Things come out of the earth, briefly, as the sun sinks further in. The earth expands and what is normally in the shadows and penumbra of light, comes out and is visible and is real.

I say out loud, to the sky, to the air, to the great level space of Illinois, "I'll see you next year Manny." And amazingly enough, I do.

To test my theory I went back the next year and Manny appeared again. Emerged more like a shadow stepping out of itself really.

So, I ask him what its like to live in the dark 364 days and 23 hours and 55 minutes a year and he says it's not so bad, he just wears two eye patches and doesn't know the difference. He asks me if what he told me about the weather last year made it in the book and I say that, yes, it will be in, but I've had to wait a year to get back to this day because this is also going to make it in the book as well. Manny nods, looks around for a few seconds, and whispers, winking his left eye, "So you got patience. Not bad for a youngster."

When I ask him about his life, it turns out that he's been alive about three and a half centuries, because darkness doesn't kill, only light. It is taking him a long time to dry up he explains, or at least that's what he's decided must be happening.

Mr. Neo Blair: A Buddhist Satva

They said that the rain was coming. Fast and hard. They said it was going to rain for days. The man on the radio told everyone to cover their window wells, to prime themselves for one damn serious and dangerous storm. You could tell that it was going to be bad, because the announcers on NPR never say, "damn." I was pretty certain I could see a dark, thick, cloud bank coming my way.

I went next door to check in with Neo Blair, my neighbor. To get his opinion of the possible storm. I usually did this since Neo could read the clouds. I always went by when I thought I really needed to find out what was going on during pending, bad weather. When you live on the Great Plains, out in the middle of the grassy prairie of Kansas, checking in with others can be a valuable thing.

Neo said, "Naw, maybe south of here a couple hundred yards, but nothing's gonna happen right here by us." Neo was roughly fifty-eight to sixty years old, but looked older. He had a slow, slight limp and a fairly hunched back. His arms writhed with veins like a pit of snakes, plus he chewed and swallowed tobacco which seemed to account for his bad cough. Always, he was impeccably sincere and was, without doubt, an absolutely incredible weather predictor. When I first moved into the area he would come by once every so often, while I was out in the yard and talk to me about the weather and just tell me his predictions. Then, without being pushy or anything, he would check back in a few days, maybe a week later and say, "Did I, or didn't I tell you it'd rain?" Over time I realized he was always right. Well today, him saying it wouldn't rain was just too much. I tried to say, "But the radio said … ." That's all the further I got. Neo was laughing, "The radio said! Ha, listen to you, you little baby, don't make me laugh."

I half-smiled, knowing I was being an idiot, showing a lack of respect, but still I replied, "Come on, you can't be serious, Neo?"

Neo looked back to the sky one more time, to ease my mind. "Nope, sorry, no rain here. The wind's not swishing enough around where we are, certainly not enough to bring on a strong change in the temperature."

"Neo, there's no way you can know that, there's just no way, Neo," I said, not really so interested in the storm any more, but wondering how, with the dark clouds nearby, he could boldly just sit back and so calmly challenge the radio and all the weather reporters' predictions. Heck, as far as I could tell the jet stream had descended and was preparing to flow directly through our town.

"What'd you come over and ask for then, huh? If you didn't want to believe me, why check? What's your point," asked Neo.

He did have a point, but still I said, "Neo, don't get me wrong. I do believe you, but no one can figure out the weather to the inch. I just can't figure out how you know so much?"

"Like I said," Neo started saying, "What'd you ask for? It just isn't going to rain on our block, it's not that big a deal really. I wish it would rain. I need some rain for my cucumber plants."

Well, I noticed that by the time we were done talking, the storm was smashing through the other end of town. I could hear the glass balls on people's weather vanes blowing up from the frequent and violent, deadly lightning bolts nearby. But yet, Neo and I were only having to contend with a light shower, well, more of a mist really, and for no more than three to four minutes. The cement didn't even get wet. I smirked at Neo while the light mist fell on us though, "Guess you weren't all right, huh, Neo?"

"Humph, was off by thirty feet. I could have sworn we wouldn't even get a drop. I must've missed something in the wind. See Harm Lewis's driveway though? Dry as dust," Neo said.

He was right, Harm's yard, next door to Neo's house, was dry as cotton. Then Neo added, "Always remember, the end of the storm has to be somewhere. Look at the edges, not the center. You have to see the space between the clouds as well as the clouds. Take heavy consideration with what the wind is doing and what it's not doing."

Marveling at Neo's smart remark, I said, "What are you, some sort of Shaman, or Monk, or maybe one of them Buddhist Satvas, Neo?"

"You mean a Buddha? Maybe, maybe I am, but I doubt it. All I know is I can read the sky as clear as day."

Personal Planet System

Each night since he was four years old, almost eleven years now, Franklin Century had begged his mother to let him go outside after dark to play under the stars. Each evening for eleven years Franklin's mother had answered, "No." Franklin knew she said this because she didn't think a boy like him needed any help putting strange thoughts into his head.

One day while Franklin was walking home from school, he noticed a bright yellow chunk of something strange alongside the street curb. Franklin picked the yellow chunk up and stuck it in his pocket. Franklin wanted all the strange thoughts and items he could get.

Later that night Franklin took the chunk of yellow out of his pocket and held it above his eyes, then focused on it real hard. He stared at it a long time, then felt it tug at his grip. Franklin let the little yellow chip go and watched as it floated up to his ceiling. Once the yellow chip hit the ceiling, it spun in circles and spewed out a collection of stars, planets and moons. Franklin had found himself an entire, although tiny, solar system.

Franklin stood up on his bed and looked at the moons and stars and planets, felt their sides, noticed their temperatures and hardness. On top of one of the planets Franklin could see little flecks of life moving around, like fruit flies. Centered amongst all the stars and moons and planets was the original beaming, bright yellow chip, just like a sun. This chip was spinning and growing little by little in size. Franklin went up to the sun and touched it, and it wasn't hot like he thought it would be, but like a vacuum. When he touched the sun, he could feel something pull all the way through his feet. It made him feel light-headed.

In the middle of this exploration Franklin heard his mother knocking and yelling at his door. "Franklin, Franklin, you open this door right this instant, you hear me!"

With the solar system spinning around in his room, and the sun tugging at this feet, Franklin knew he would be in trouble if his mother came in now. So he tried to clear the air as if the stars and planets were cigar smoke, but that didn't work. Next he tried turning out the lights to hide the planets and moons, but they only started glowing brighter.

"What do you want, Mom?" he asked.

"IN!" she said.

Not knowing what to do, Franklin stood on his desk top and pushed the sun into his mouth, then swallowed. As soon as Franklin did this all the stars, moons, asteroids and planets came roaring into his body, seeping through his skin, gushing past his ears and mouth. It all took about fifteen seconds. Plopping down on his bed, Franklin said, "Come on in, Mom."

Franklin's mother looked in and stood for a bit, said a variety of things, ending with, "It's time for bed now."

"OK." Franklin said. At that point Franklin's mom should have known something was a little wrong because it was the first night in eleven years Franklin hadn't asked to go outside and play.

The whole time Franklin's mom was talking to him, he was seeing stars drift along the edges of his vision. When his mom left the doorway, Franklin found that he couldn't breathe easily.

Coughing, he spewed the whole solar system back out into his room. At least everything but the sun, which seemed to have stayed inside him. In fact, he could feel a small fire, not painful, but warm, deep in his stomach and sunlight was flowing off his fingertips in beams.

Soon Franklin's life became like something from outer space. If he opened his mouth, stars came out. If he closed his eyes, he felt like he was floating through the ozone layer. It was hard to be around others and so he started staying home from school. He had to begin eating through a straw and needed to learn to chew with his mouth closed so the food wouldn't float back out.

Then he started changing in other ways. He was beginning to defy gravity. When he sat at his desk at school, the table hovered about half an inch off the ground. Next he saw his body shape start to change. His feet and his hands were becoming round orbs, his toes and fingers were merging together.

Franklin discovered his need for air was diminishing, it wasn't long before he needed just one breath to get through an entire day. Slowly his weight dropped, and just to stay on the ground, to keep from floating away, Franklin had to tie a big bowling ball to each of his ankles, but even so, on windy days, he felt he might blow away.

Then he started to look dirty. The last day Franklin went to school was when his friend Pete asked him, "Hey Franklin, what's that cloud of stuff around your head." Well, even though it was obvious, Franklin had hoped no one would notice. There was a ring of meteor dust around his neck, up behind his head. Scared of what would happen next, Franklin began telling his mother he was too sick to go to school.

Two days after dropping out of school, Franklin began having to tie steel cables to his arms and legs and would float outside his home like a blimp in the sky. The round lumps that had been forming on his feet and hands had formed into balls. Finally, one afternoon these lumps simply popped off his body and began orbiting around Franklin's head. He was slowly becoming an entire solar system. He knew things were completely out of control when he could feel a nighttime sky on one side of his body and daylight on the other.

He also knew he was in trouble when his mother came home and saw him tied up and floating, like a kite, in the sky. "Franklin, get your butt down here NOW!" she yelled.

"I can't, Mom. I can't." Franklin said.

His mom just turned her back and walked inside. She was muttering, "Some strange thoughts have sure taken over that boy."

That night, when Franklin had managed to pull himself back down to the ground and anchored himself at the dinner table, he heard what he had been waiting years to hear. He asked his mom if he could go outside and play under the stars and the moon. Unbelievably, his mom said he could.

Franklin smiled, thanked his mother, and realizing that this was the end, sighed with relief. He knew there was no hiding it anymore. He understood that he had grown up and needed to leave home and move up to the sky where he belonged. He gave his mom a quick hug, went outside and rose slowly up into the sky.

Vanilla Ice Cream

It is an old day. A long time past, but the memory is bright. It is the color and thickness of fresh dairy cream during summer.

The day was hot and humid. The air was alive with summer crickets, frogs and birds. It was a land, a time, a place that brought us all together. A place no longer possible. A place to revisit in the mind.

The sun is never going to set on this hot, sunny, central Texas day. The whole family is gathered, grandparents, aunts, uncles, friends, cousins, dogs, horses, cattle, deer. Everyone is welcome. It is deer hunting and ranching territory, but not today. Today we are all making vanilla ice cream. The guns are resting after a few hours of trap shooting. Irene is sitting back drinking iced tea. Annie is talking to Bernice, and the other Bernice is talking to Warner, who is standing next to Jewel. Bud is just now telling someone that he's going into the trailer to listen to the Astros game on the radio. Two other people say they will join him. Coming toward me and my brother is Gordon, and we are waiting. We are standing restless. We know that we are being raised to be silent, polite kids, not to make scenes. Never heard, just seen. We are supposed to be good little boys, but today we quit pretending that we believe in these roles. After all, we will be making ice cream. We both know this is special, and so far in our lives of six and seven years, we have never made ice cream. So we want to start cranking, we want to get started, we know it will take a long time to whip up the custard that we saw my aunt pour in the ice cream maker, so we are anxious. We are in a hurry. We want to get started. We want to make the ice cream ... Come on, come on, come on, hurry hurry hurry ... We can see the ice cream maker in front of us. We yell and hop around in short tense little steps, whirl around, running in place trying valiantly to make things move faster, "Hurry up Uncle Gordon, Hurry up!" We finally yell.

Meanwhile, more people are arriving. It is incredible. Usually all these people don't get together. Frank and his wife are driving in. His driving style is quick and then slow, not too certain in close confines. He likes to scare people, so his driving is as much a joke as not. I always wonder if his foot moves on the gas pedal by choice. I believe he can't control his pedal foot too well.

There are also Joe and Jimmie, and as usual they are in a good mood and there is a small group around them. Joe is showing someone something from the trunk of the car, an engine of some kind, and behind them, in a home-built trailer, is his brother Dick and his wife Herma. I understand that tomorrow we will have fresh oysters and shrimp that Dick has carted here in his trailer from the gulf coast. Babe and Allen have just driven in with a spanking, bright, new, silver Cadillac, and this gets everyone's attention for a moment. As they get out of the car, everyone says "Nice car," and Allen says to everyone that he got a speeding ticket, and Babe laughs a little, then clutches her bright, neon colored wicker purse and walks to the drinks with a cigarette in her hand. I notice she wears gold-colored shoes and that Allen is wearing a suit and hat. Very different dress I think. Driving in next are more folks, the ones from South Gulch, all twelve of them: three boys and three girls accompanied by six adults.

There is even the lady we know as Kraft plodding in, and I am a bit worried. I'm hoping she doesn't see my brother and me because then we'll have to go over and do the bear hug routine. Next to her are some more relatives named Clarence and Anne and two other people by the last name of Green, alongside them. This group is mostly named McDuff. They are all sitting pretty still. They are all holding court around Clarence and Anne, as they are the oldest people here on this day and, as they always do, they are getting everyone to sign a guest book. Then as people sign their names, Clarence stares at the signatures, noticing how they've changed since last year, or from the way he remembers them having been in the past. He is busy discussing writing styles and legibility. I think to myself how interesting it would be to see the whole collection of signatures and hear what they've noticed, what they think has changed.

We are all surprised to see that even my Dad's parents have arrived with their favorite dog, a brown-and-white bird hound. Everyone is glad to see them. There is another small clustering of folks as well, people I don't know much about, and who at first I hadn't even noticed. They are off under a separate tree, all standing around some important relative named Carl. Everyone is always so hush-hush around and about him. My eyes cross this group, but move on. I am not interested in the hush-hushness. I understand he is rich. I wonder if he'll even want any ice cream. I doubt it.

I recognize many others, but there are so many, and after awhile I don't care about identifying them because Uncle Gordon has my right arm in his hand, and he's showing me how to crank the lever around and around, how to add the salt, and so I start to do so, and as I begin I look up and he nods, yes that's it, that's good. So my brother and I turn the ice cream crank and add salt in turns, and look around and people ask how much longer, and we smile and say not much longer now. Then one of my relatives comes by and turns the crank a couple times, laughs, then says, "You've got a long ways to go son." We keep at it, looking around at everyone, until finally, when the shadows are starting to lengthen into evening, after the food from dinner has been cleaned up, Uncle Gordon checks the now-hard-to-turn crank and says, "That's good. Let's eat."

We watch him lift out the cranks and pulleys and whipping fins. There inside is the smooth, vanilla ice cream. We lick the beater blades, and it is good and we smile.

All these people being here creates a feeling of being part of a deep and sincere family. It is good and it is necessary to feel so rooted, to have a memory of a place in humanity that doesn't involve any strangers. When I think about all the relatives I have known, all the folks that have wished me luck, I know that I am lucky, even if most of them have passed on. But back then, when I was small, I was able to know my past, and I can't help but carry parts of them all into the future.

So this is my memory, all these folks sitting in the summer air, under the shadows of the oak trees sharing their stories with ease. They were all so comfortable that they were able to keep telling the same stories again and again, because they could. Everyone wanted to hear the good parts of life again. What modern doctors might describe as senility is really a celebration. Every once in awhile I heard something for the first time at a large family gathering, but usually I heard things for a third or fourth time, yet I always learned something new. For one thing, I learned that what was really important was what people enjoyed repeating and hearing over and over. When the good parts of a story were coming I'd look at my brother and we'd grin, knowing, "Here it comes, remember?" It felt good to be comfortable with these stories. Then we would take over another scoop of ice cream to an aunt or an uncle. And when they took a bite, we would look into their faces and they'd wink, or nod, squeeze our shoulders, indicating the words, "Thank you."

It was even better when one of my grandparents, or parents would tell a story. When I could look at my grandfather across the group, his hat tipped up, a beer in his hand, his face tanned, begin to tell a story I liked, I would beam and wait with anticipation, watch his eyes brighten up until finally, everyone would laugh. His face would look rewarded with everyone's laughter. It was always a good feeling to be able to look around and be proud, to know it was my grandfather that just said that.

These gatherings always made me feel completely welcome and part of the entire world. Everything made perfect sense and everyone was safe and it seemed like everyone always would be safe and alive and right forever. It gave me a feeling that all my decisions would always be right during my life.

Now I am beginning to realize that this is what real life is about. Collecting the good stuff together a few days a year and being able to smile in a group without doubt. That life is about storing away good memories that give you a sense of time and community and pride.

Those early days also let me know something else. I learned that when the ice cream was done, that when everyone congratulated me and my brother for a job well done, when everyone was eating and looking around, that you didn't have to tell everyone you loved them, because that had been taken care of in the making and eating of the vanilla ice cream.

SPONGE FOOD

It was strange at the time. You see, it had never snowed in my hometown of Rocknell. So when my friend Wilty and I returned after a week-long skiing trip from way up north, we thought it would be fun to bring some snow home with us. We were going to show it off during the summer.

It wasn't until later that we found out the problems, me especially. We thought we were just putting snow in a freezer. Before the last week of school we said to our teacher Mr. Frowncult, "We got some snow to show if you'd let us."

Mr. Frowncult looked surprised and started talking, "I'm sorry boys, but it's summer now. Let's not hear talk like that." He answered us in a smug tone.

We reassured him that what we had wasn't a joke. That we had snow in my freezer at home and if he would let us, we'd bring it tomorrow to show off. Mr. Frowncult locked his eyes open and went home sick after we said this.

I asked Wilty, "What did we do?" Wilty didn't know, but something out of the ordinary seemed to have sent Mr. Frowncult home that day. He had felt a giant lump forming alongside his neck, in his throat. He'd been unable to even walk straight and had missed the rest of the school year. We never did bring the snow to school.

When the school bell rang for the last time that school year, Wilty and me instantly forgot about Mr. Frowncult and looked forward to some serious playing. We planned on celebrating the beginning of the summer with a small snowball fight behind my house. We had some bets to take care of first, though. Heather and Lewis Storage had never seen snow before and said they would pay Wilty and me a dollar each to see the snow we had. At least they said they would give us a dollar if we could prove to them that what we had was snow. We didn't know any better way to prove we had snow than to show them the snow.

"Can't have no snow in the summer, that's all I know, can't do it," Heather told us. We knew she was wrong.

It was weird, though. We showed Heather and Lewis our snow, and they called us liars. We told them there was no way we could make up something like snow, we just couldn't. We asked them, "What do you think we did, freeze cotton?"

Heather just shouted, "Liars," again and again and started crying. We saw that Lewis had turned just like Mr. Frowncult. His eyes were locked open and his throat had a big lump on it.

He swallowed hard and looked like he bit his tongue.

"Liars liars liars......," Heather kept saying, louder and louder, "You can't have hot and cold together, Y O U A R E L I A R S!" Heather and Lewis didn't give us our dollars. They went screaming, back to their house.

Wilty and I both agreed. We had never heard anyone yell as loud or as worried as Heather had. Wilty began to wonder a little about our snow. To ease his mind, we put the snow back in my mom's freezer without having the snowball fight. Wilty said he felt a little uneasy about the snow suddenly.

A little while later, Wilty asked me a question. "Do you know what *real* snow looks like?"

I looked at Wilty and said, "Yeah, just like what we got." Wilty wasn't satisfied though, I could tell.

"I mean what *real* snow looks like," he said, "I'm not sure what we have is real. People are right, you can't have hot and cold together."

After Wilty said this, we didn't play much more that day. Wilty reminded me too much of a grown up. He was being tough minded and not making any sense. He seemed unable to respond to new ideas. He seemed more inclined to simply agree with others than to believe in what he knew was right.

I didn't want to believe that my friend Wilty was as bad as a grown up though. I wasn't very old yet, but I already knew that being a kid was better than being an adult would ever be and so I wanted to be as much a kid as possible, even after I got old. I could already tell which adults had hated being kids because they never joked, or listened to music, or played, or moved faster than a walk, plus their voices were boring–always the same tone.

I was bummed out. Wilty knew the truth about the snow, and he still wouldn't believe it. So, as Wilty walked off that day, after questioning our snow's reality, I wasn't even paying attention to him any more. I was humming a song and carving something with my knife.

The last thing I remembered hearing Wilty say was, "I guess we might as well give up on the snow. It's probably not real anyway. I guess we got to grow up some day."

I didn't know what Wilty was talking about and didn't want to ask him either. I called him up later that night to ask though. I said, "Wilty, what were you talking about when you said that thing about being grown up?"

Wilty said he couldn't talk right then, that he was busy, and hung up on me. Things were getting to be secrets and hard like rocks. I was getting mad. Wilty was being a grown-up at age twelve, just like everyone else. For no reason he was being single minded and sick, like a lump of cold, boring, sand.

I didn't see any more of Wilty for awhile. I finally saw him again at the City Balloon Fest, where everyone ties their name to a balloon and sets it free at once. Then there's music, food, rides and stuff. Wilty and I matched eyes and then Wilty walked over and asked me about the snow. "I still got it," I said. I asked him that thing about being grown up again.

Wilty said, "It's nothing. Might as well get rid of that snow stuff." Then he went off toward his mother. He even ran like an adult over to her, all uncoordinated with stiff arms. Wilty wasn't much to me any more. I decided that he wanted to be a sponge when he grew up.

Later that night, after having seen Wilty, I sneaked past my parents, who were sleeping, and got the snow out of the freezer. I carried it outside. The night was filled with stars and the moon was shining bright. I held the snow balls above my head and looked through them. The snow glowed in the moonlight. The snow balls were almost see-through and had become crystal balls. They began to give me visions. I saw kids playing in the sky, people swinging from vines, dogs running. I spotted whales swimming in the ocean. After holding the snow balls up to the moon I got an idea.

I knew it was mean when I thought of it. I was going to steal Wilty's chance of being a kid from him. I was going to do it real obviously so he wouldn't think about it until it was too late. In some ways I didn't want to do it, but Wilty seemed to be asking for it.

I invited him over to my house. I showed him the snowballs and then put them in glasses. I let them melt. Then we drank the water. We swallowed hard when we drank. I was drinking to keep my secret. Wilty was drinking his dreams away.

"See, it wasn't snow after all. It was just plain old normal water." Wilty said.

"Yep, can't have two things at once. That's for sure," I lied.

Then Wilty said just what I knew he would, "It sure is weird not to feel like a kid any more."

"Guess so," I said.

When Wilty left, I sat for a few moments and realized that when I grew up, I was going to be brave and bold, not stale and frightened. I was going to live alongside excitement and new ideas no matter what. I wasn't going to give up my imagination, or give up playing games, or having fun. I didn't want to be a sponge and only be able to soak up awkward ideas about how and who to be according to other people. I wanted to try and figure out mysteries and not be scared of things that were different. I knew that I would never run from what I believed in.

Name Your Thunder

Depending on how personally you take your weather, your rainstorms, your thunder claps and your wind, you may want to consider naming the moments that you can remember, and better yet, begin naming storms that you know are on their way. A hurricane is partly named because it will soon be getting personally involved, very involved, in the lives of those in its path. It will be "alive" long enough to be treated on a first name basis. I can think of many storms that have passed over my life that I still remember with clarity.

Many inland-weather occasions take on very personal forms. Perhaps your front window has been blown in, the outside of your house dented and destroyed by hail, or some of your trees simply twisted down in a crunch to the ground during an ice storm. If any of these events bring back a memory, then you know that the idea of naming for honor and memories sake isn't strange.

To get a glimpse of the world of names, journey with me to a land of language, to a little world of stories. To where the collective consciousness of a small town and many of its generations reside together to share the past. Listening to the gathering of people is to hear a shared knowledge. Journey to my area of the world, to the foot of the Iowa Mountain. Follow to within earshot of the group we call the Higher Than The Sky Gang. It is from these people that the young can learn where names and language are kept alive.

Higher Than The Sky Gang

Enter: Mid-morning, a cafe, dishes clattering, and in the back corner of the old wooden walled cafe is the noise of muffled chatter.

Some time long ago, unexpectedly, probably long before stories had even been invented, men started to gather at the Bitterhaus Cafe to compare life and all that their lives had meant. Within these thoughts were glimpses into heaven and tiny holes drilled to hell. Walking in through the front door now is more clutter to add to the cluster of men. It is Thelbert Sheridan, and he looks to have something on his mind.

Thelbert is a large, gray-bearded man of about eight and half decades. If someone were needing a new cologne they could use the smell of Thelbert—call it Rugged Age. As he sits down, he's talking to himself as much as to those around him. He is mumbling something about the suit coat he ate out of desperation during the drought he had named, like coastal people name hurricanes, Bankruptcy. Everyone nods in memory. They all remember that summer. He is talking about how he had dressed up fancy to salute the rain he hoped would fall. Like a bad joke, he had begun to make up rain dances that summer. He was becoming weird and like a wild thing, and he knew it.

He also realized that he was becoming alive at the same time and hoped that rain might come of it. He'd dressed his body up with good clothes, slicked his hair extra well and even written a prayer to the sky. He read the prayer in the most isolated tip of his corn field. Out in the corner of his life where no one would ever think to go, or yell at him to do anything: *"Oh great sky, so high and above me, sky all around me, sky which holds water, sky, Oh Sky, please drop rain. Please let yourself fall to the ground, leave green colors at my doorstep, break back open the creeks that have forgotten how to flow with water, put my seeds to use, give me corn, leave me life. Give part of your sky up soon, Amen."* Thelbert said he tried this prayer and many others in this corner of his field that summer, out in the spot where he could cry and never be discovered. He had survived the drought of Bankruptcy, but he says he lost his faith in God that summer. Lost his faith in all but desperation and the power of nature. He became enchanted with how people can twist and create a world of internal adventures deep inside their head. He wasn't proud, but he had switched from God to Clouds for a sense of faith. This is because he wanted to believe in the most powerful thing he could think of, and that was definitely no longer the human creation of God. After he said this, he sat still. He didn't really look at anyone else either. He sat a bit ashamed, but well convicted of his mind.

The littlest man in the group, Grit Johns, starts in next. He opens his mouth while nodding at Thelbert's talk of clouds. Right as Grit starts to open his mouth, Jack Kerns cuts in, "Not another of them mountain top tall tales now, Grit." (everyone snorts and laughs lightly, but, all the same, wait for what Grit has to say). And Grit does go on talking. He tells about how he was friends with an old man from over by Buck Creek when he was little. How that man had spent two months eating his electric fence. That he had taken a metal file and made a pile of electric dust which he added to his coffee each day. Then after he'd eaten the whole fence, he roamed out to the prairie and turned himself into a lightning storm. He would light up and howl, yelling like thunder. Grit said he really did sound enough like thunder to curdle milk, by scaring the cows.

The gang of people all smile. One person says, "I think I heard about that," but he says it real soft, half serious, half joking, not wanting to commit too much either way, for or against Grit's story. Believing the far fetched always seems to make men uncomfortable.

Another man, Fleck Peters, just rocks back and forth in his chair, picking at his teeth. Fleck's simple and slow smart. He listens well. Rocks.

Picks. Nods. About the only way you can tell for certain what Fleck's doing, whether he's listening or not, is by following the gaze of his eyes. Fleck pretty much seems to determine who talks in some ways. Seems that wherever his eyes stop and wherever he spends time gazing, talk usually picks up. Slowly now, Fleck moves his eyes over to Charlie Penny. Fleck is stare-asking to hear about how Charlie's mother used to have puppet shows with moonlight.

Charlie feels the glare of Fleck and winces his face and says, "Oh no no no, not again Fleck. You want my Momma's puppets more than me." Charlie then holds his hands up to the light bulb and casts his mother's imitation of an African Hippo on the ceiling. Although an interesting image, Charlie does this with such little emotion that it is sad to see. Charlie rarely looks forward anymore, he is too busy looking down at the ground all the time. It turns out that Charlie doesn't remember how to laugh anymore. It was three years ago that his mother died and since then Charlie hasn't cracked a chuckle or even split a grin. Behind Charlie's back everyone is agreed that he's acting kind of like a baby, but no one would ever say that to him, not here in the realm of the Iowa Mountain. To do so would just be too crass, plus, everyone agrees, in some ways, to still be acting like a child at age seventy-four is nearly admirable. At the same time, when your mother dies you shouldn't get mad and depressed about it. Everyone's mother dies at some point, most before their children do. Charlie still carries a five by seven inch glossy picture of his mother, framed, wherever he goes. He goes so far as to set the picture on the Bitterhaus Cafe table when he sits down. It's agreed that he's strangely alone and homesick. Looking at Charlie, though, it does make a certain amount of sense, his sadness. He is truly nothing without his mother. Since she died he has slumped something awful. He's literally become ugly, tattered and sad. He looks like a gray piece of creased skin. Some of the people at the Bitterhaus have begun to look away from Charlie when they see him, "Just too sad, too sad to see," they say.

There's more to know about Charlie though. Hank Ferguson tells it best. He remembers that Charlie hasn't really laughed for fifty-five, maybe sixty years. Not since his inside and face were burnt out. What happened was that the first year Charlie had been out of Hercules High School, he got a job at the filling station. He had been checking the oil in a customer's engine when the motor just flared up. The flame burned through Charlie's face and spread right down into his lungs, going so far as to puncture his stomach and heart. Since that day Charlie hasn't had the breath or desire to smile or laugh, at least not so another person could tell. At the same time, there's no arguing that his mother's death took the last glint of humor out of him. Those that know him best say they can still recognize when he tries to smile, and profess that he does so from time to time. Yet it's safe to say that Charlie doesn't have to worry about being hurt by laughter anymore since the chance he's going to find anything humorous is next to nil. As Kenny Ozark says, "The day that boy laughs is going to be bittersweet at best. I'm afraid there ain't nothing much a human could do about getting him to grin again."

There may be sad stories, but nothing ceases the talk completely in the cafe. No matter what, in this room memories always keep pouring out. Nothing stops the half-remembered lives from being crossed up, tossed around and stretched between the semicircle of folding chairs, donuts, coffee, ash trays and piles of smudgy newspapers. Looking closely, a clue about the majority of the conversation stands out. The newspapers are all opened to the baseball score cards. The reason is slow to understand, but worth figuring out. The solution comes about as all things do in this room, by listening, waiting and sitting.

Eventually it becomes overly obvious that everyone sitting in the Bitterhaus Cafe has played some low-level, pro ball at one time or another. It's a bit strange that no one ever played against each other, yet still, they know one another by the nick names they left behind: Plug Spit Ramboline, Not-A-Chance Pete Peters, Round Mound Randall, Chickenboy Himpel, Curtis Long Ball Campbell, T. Telly Hutton, L.B. Maxwell, Cobra Weno, Gino Geronimo Julio Chavez Suggs and on and on.

In fact, Mammoth Spin McGill is talking to Dave Very-Red-Man-Chew Davis right now. They are remembering when the Howard City Flyer's Manager, Slim-Feast Williams, spit a hurricane of tobacco juice in the face of plate umpire, Dean Beefer Rollins, for calling Lemon Drop Jackson out during the deciding game of the playoffs. Instantly, memory kicks in and talk goes like it did that very day forty years ago: "He was out by a mile!" chimes in Lucky Roy Schmidt. "Oh no he wasn't! He was safe by the moon to the sun," shouts out another. It's clear that it takes a long time for some things to cure in old minds as well as in the young. In fact, some things are probably never forgotten, or forgiven. Some things are more interesting as arguments. To forget, or forgive would be to give up on the fun.

Then people start talking about how Manager Twinkle Toes Jacobsen used to call for rain-outs in the middle of sunny days to try and avoid losing a game. Once he threw a bucket of water at the umpire to try and be convincing, "There, soaking wet now, ain't it?" he used to say, grinning.

Then someone mentions plate umpire Jules Stanker DeQueen. Jules had been so proud of his strike call that he used it to try and show off to the ladies in the crowd. It took him ten seconds to call a first strike, twelve seconds to call a second strike and about fifteen seconds to call a third strike. The third strike, in the final inning would last until the final fan left the field. Someone mentions that ol' Jules never did get married and laughs, adding, or even get a date. This triggers off more laughter in the Bitterhaus. Someone mentions the time that field umpire Clem Stare Houston stopped the game just to watch the clouds pass by. "Damn good sight to behold," he'd declared. "Nothing like a good sunset during this life as we know it."

When Charlie Penny hears this recollection about the sun and Clem Stare Houston, his face winces up hard and he looks to be swallowing a boulder. A couple of guys guess that Charlie may have even cracked a smile, "Naww," they both agree, "Impossible."

When Francis Chicken-Bark Parks sees lipless Charlie swallow hard, he twitches his chin akimbo. Francis has loose nerves, so after he looks at Charlie, he moves his chin, quick, down to his chest, trying to act like he wasn't really looking when Charlie looks him in the eyes. Francis turns a bit red when he knows he's been caught staring, so he moves his gaze out the window and begins to chew on the gum built into his lower lip. He watches the birds in the trees and the sky above. It isn't until Francis sees a blue bird against a blue sky that he is able to feel his heart begin to slow down.

For some reason staring at the sky and the birds is comforting to Francis. Francis doesn't do a lot of talking at the Bitterhaus. He just likes to be alone, next to everyone else, as if sitting alone in the backseat of a moving car on a cold winter night, all warm and insulated in the dark. But one time Francis did tell Charlie why he liked the sky. When he looked up into the sky, he was able to imagine what it was going to be like when he died and moved up there to live. Francis said he thought it would be like sleeping on a pillow. He even admitted, a bit shy, that he could hardly wait to pass on and quit feeling so useless now that he was so old and alone. He hated being alone, stuck with a bunch of memories he didn't like getting in the first place. He said he didn't like talking and thinking about the past, but all the same, he knew that's all he could do. For the most part, Francis wondered where his life had gone, "My damn life ain't fit for shit, never was, but heaven sure will be," Francis would say. Then he'd place his chin back down by his chest, or look out the window on sunny days. He was often heard muttering, "it's seems later than I thought today, much later than I thought."

On the other side of the room Checker Board Willis is telling anyone that will listen that he had bet on horses in Viet Nam and won a million dollars on a pure breed named Max Giddy and that he had slept on the peak of Mount Everest for two days. That he'd built sand castles in Korea under a full moon during gunfire, been killed three times in a car wreck with a cat in his arms, cursed toe-to-toe with Lucifer and won, heck, he'd even sold counterfeit money to the U.S.

Department of Treasury wearing a communist uniform.

After talking like this for twenty minutes, Checker Board Willis nods his head, smiles and says one more thing before he falls asleep. He declares that he has purchased two hundred shares of stock in the promised land.

It isn't long after Checker begins to snore that just about all the men in the room nod their heads down and say they have bought stock in the promised land too.

Soon everyone in the Bitterhaus Cafe is asleep. It sounds like clouds moving in the sky, the way everyone's breathing so deep.

Orange Blossom

It is twenty minutes before sunrise during a late, orange leaf, October morning. The crest of the earth out in the distance is bright with various wisps, sprays and tendrils of umber colored, glowing sunlight. These forms of deep red and orange hues will place a harvest moon in the sky tonight. This holy light is stretched over the entire township, here in Iowa, just west of the

Mississippi. I imagine this softer, slow and steady morning light to be a gentle form of the Northern Lights, perhaps it could be called the Midwestern Lights–so many flashes of brilliant, earth toned, orange color in the eastern sky. The light looms trapped in cold stiff grass, surrounded by the call of turkey, enhanced by the fermented smell of apples on the ground. The lights of morning are the petals of an orange daylily blossom.

On this morning I hold my cat up to the kitchen window and he meows and then jumps out of my arms, going over to stand by the door. I open the door and out we go. I ponder my interest in the faint, early glow of the sun. Wonder, why am I so constantly attracted to soft light. A light more like a drift than a shine. I believe it is because dim and faint illumination is able to squeeze water from rocks, from air, from the land. As a new day begins the world is freshly created, all covered with a coating of fresh squeezed dew. One has to view each day in amazement. Has to marvel at how the sun and the weather and the world is continuously successful at invention, at originality, at preservation, at moving and changing without once lapsing into replication. I wonder, as I watch my cat paw in the dirt and eat grass with contentment, why I have such a hard time just living and accepting the world as it is. My cat looks at me right when I think this and he meows, soaking up the sunlight in full and comfortable satisfaction. I know the world makes perfect sense. I seem to think too hard to believe this.

Plastic Puddle Drops

The beginnings were with my friend Lenny Rim. Lenny told me a secret his Uncle Wendell had been telling him every summer for the last five years. I guess it wasn't the best kept secret around with everyone telling each other, but shortly after I heard Lenny tell me what the secret was, it did seem worth hiding from most people.

The secret had to do with logging and it was the kind of secret that began to knock things over. Lenny said it was only natural for his Uncle to tell him something about logging since he had lived up in northern Minnesota all his life, cutting trees for money. Lenny said his Uncle Wendell was like a bull: sturdy, stiff and silent. All Wendell knew about, or cared to know about, was logging. Which was fine, since every one of his friends had been loggers as well. Real quickly, before he told me all the details, Lenny said the secret had to do with rain drops, puddles and trees. Lenny also said that the secret worked just about in that order too. Once I found out about the secret, I added my own part at the end. This involved plastic food wrap which had to be shaped just right with the palm of a hand.

Lenny told me the secret over a couple of days while we were working at our combination lemonade/treehouse/snack shack which we had built in a vacant lot between our houses. It was placed alongside the road so cars could stop and order without the passenger even having to get out. We sold all sorts of things to people, mainly peanut clusters, lemonade, and white chocolate covered pretzels. Usually we made out all right earning about twenty five dollars a week, each. We often wondered how we ever made any money since we never really did anything but play around and talk. On a piece of paper inside the store we had a top five customer list that we kept updating. Pretty much the same five people were always on it, they just changed in order from time to time. The first on our list of best customers was Eric Hollery who delivered the newspaper in the neighborhood. Basically he was number one because he always bought something. We didn't really like Eric that much, but we did like his money, so in the end we both agreed that he wasn't too bad a guy.

The second person on our list was Mark Eds. Mark was a college student that walked by twice a day and was just a cool guy. He was laid back and not like other old people because he still ate candy, played catch with us and gave us his baseball magazine each week. The only real problem with Mark Eds was that he didn't really buy anything. It seemed like he had a new girlfriend every week and they usually bought something from us when they walked by though. Also, his girlfriends were as interesting to see as the baseball magazine. I suppose that if we would have told anyone else Wendell's secret it would have been Mark Eds.

The third person on our list was Paula Clark who delivered the mail. She was quiet and nice and predictable. Plus, we had to see her practically every day, so she gave us something to look forward to.

The last two people on our list of top five customers were Aaron Deeder and Layla Sims. Aaron was a tall kid with a rich dad and a poor mom. He lived with his mom, but every other weekend he saw his dad. So, every other weekend he had loads of money and when he had money he bought our white chocolate pretzels by the handful. He considered them the best things he'd ever eaten. As for Layla, she was my secret girlfriend. Lenny liked her too, mainly because he liked watching me get nervous when she came by. Since I wanted her to like me I basically gave her five times more than she paid for. I told her that it was a special deal we had going. She was five years older than I was, so she probably knew what was going on, but she kept stopping by the stand so I didn't mind if she knew or not.

Well, it was between waiting for customers, like the ones I mentioned, that Lenny told me his Uncle Wendell's secret. He said that in Minnesota the kids learned about logging by dancing with rain drops as they bounced into and off the ground during rain storms. In the spring, the real lessons of school were held outside on the playground. Some of the older teachers and a few retired loggers would volunteer to show the kids how to dance with the rain. Most of the adults didn't want any part of this water dancing though, they thought it was nonsense. It was a hidden lesson that was carried on as much through superstition as fact, and definitely not taught inside the walls of the school.

Of course, most of the kids weren't interested anyway, or had been told by their parents not to listen. Unfortunately, the only real chance you had to learn about water dancing though, was as a kid. It was the type of secret that could only be passed on by faith. Even then, many kids who had learned water dancing early on in life, forgot since they gave up believing. As most people get older they get entrenched in the ways of peer pressure and fads, becoming more comfortable and fluent with simple facts and losing their dreams of possibility, hope and imagination.

In Minnesota it wasn't called dancing though, it was called Water Logging. It was during Water Logging that the trees in Minnesota grew the fastest. Supposedly, when a person learned to Water Log properly they could look into the forest and watch the growth of trees. Water Logging allowed a person to watch the bark on a tree stretch and wrap around the trunk. One could see the color green being mixed together with soil and water and sun. Lenny said Uncle Wendell explained to him that learning to Water Log was easier than things like reading and math. Wendell said that when learning could be done while running and thinking at the same time, it was much more valuable and useful. He thought math and reading were a waste of time compared with learning how to Water Log. Me and Lenny agreed, we didn't have any problem with this theory of learning.

The next part of the secret was something Wendell called Magnitude. This began a little after it started raining, while everyone was Water Logging. Magnitude was when the rain drops got to be larger than life. Lenny paused while he told me this and said that when his Uncle Wendell described this part of the secret, he would stretch his arms out broad and loop them wide around his head. These arm movements were supposed to emphasize that the drops of rain were like wild things, covering everyone in a sort of serum that made your mind rattle and boom. The drops of rain became larger than anything the mind could fathom. Wendell said that when the rain drops got Magnitude, you felt like you were slowly being squeezed between the fingertips of a giant. Magnitude was a thing so large that everyone had to raise their arms up high above their heads to become as thin as possible to avoid being squeezed to death. After awhile the Magnitude got so large and strong that people were punched and popped, up above the rain drops. This was the start of the next stage, when you were standing on top of the drops of water. At this point the water became, what Wendell called, Puddle Drops.

Standing on top of the Puddle Drops was like being on an oversized fish egg, or a chunk of gelatin. Wendell knew this probably sounded like a weird thing, but he said that when you were actually standing up above the rain drops it wasn't strange at all, in fact it made perfect sense. It was the next thing that happened which really surprised people: when the rain slowed down and the trees started drinking. People up North called this tree drinking Water Buffaloing.

Water Buffaloing was the loudest noise of all. Herds and herds of Puddle Drops were pulled through giant yellow birch, white birch, pine trees and oaks, like rapids on a river, like huge carp sucking at the edge of a pond. Water Buffaloing was a lot what flying would be like, Wendell said, because when the Puddle Drop you were standing on top of got sucked into a tree, you were pulled in as well. This pull through the tree was like a dried bean shaking in a tin can. Being pulled through wood was important because being inside the tree was where you learned to be a logger. Inside the tree you could see the grains of the wood and how limbs grew out from the trunk. Being inside a tree never lasted very long, maybe five or ten seconds, and then you'd be squeezed out a sprouting leaf. Like Wendell had said, this sort of learning usually held no point for adults, only kids. You either learned about Water Buffaloing young, or not at all. This presented a certain sort of predicament Wendell said, because sometimes you can only learn things when you don't want to, and when you need to know something, you can't learn it anymore. This was the whole secret according to Lenny.

After I heard the whole story I wanted to try it. In fact I started looking up in the sky for rain right away. I'd have jumped for joy to see a deep, dark cloud moving in. I wondered, of course, why Lenny had never tried Water Logging, or Buffaloing before. He said it was because he figured that Wendell was probably pulling his leg and he didn't want to make a fool of himself. This didn't sound like much of a reason for not trying to me though. Lenny, however, didn't dare do something he might be laughed at for. He had always been an easy target for jokes since he was rather thin and easily embarrassed. He was scared stiff of being different. To his credit though, he had a way of living through other's actions that gave him a sense of accomplishment.

I told Lenny I'd try everything out first and tell him if it were all true. This way Lenny could laugh at me if everything turned out to be a big joke. I could hardly wait to try the secret out. I was doing rain dances in my head.

It was a week before it rained, but during that time I came up with an idea of my own to add to Wendell's secret.

What I'd been thinking made perfect sense to me. I was thinking that there was probably a way to wrap trees and puddles and drops into some sort of container. Then you could have a Water Log in a jar, a Water Buffalo in a bag, or a Puddle Drop in a pouch. I wanted a piece of every part of the secret in my room. This way I'd have the chance to experience Magnitude whenever I wanted. I told Lenny what I wanted to do and he sort of chuckled. He said it would be kind of hard to do since we didn't really know much about Water Logging, or Water Buffaloing in the first place. I ignored Lenny of course and figured out my part of the secret.

It didn't take me long to think up. The container I would use to trap the secret would be plastic food wrap. Lenny asked me if this was like trapping a tornado in a soda bottle. I didn't know for certain what Lenny was talking about, but I told him, "Yep, it's just like that."

When it finally started raining I raced outside. Flying out the door with my plastic food wrap I did what I had decided to do. I stood next to our big old sycamore tree, dug a three foot hole and then smoothed the plastic food wrap out wide and taut, carefully with the palm of my hand. I was trying to hide the plastic from the rain and trap it. All in all, I thought I'd come up with a pretty good plan.

Soon the rain was coming down like rows and rows of men playing Russian roulette. I was only partly nervous and a whole lot scared. I watched the drops of rain bounce up and down on the plastic food wrap and start collecting into a puddle. I was standing alongside, watching, when I began to bounce up and down, off the ground. Slowly, I found myself raising my arms above my head. I was having a hard time breathing and completely forgot about my trap. What I'd done was go straight into Water Buffaloing. I could tell because my chest was being crushed and I felt as if I were hovering over the earth, like a cold morning fog.

Everything seemed over before it started. Had I been fooled? I guess I thought flying through the trunk of a tree would be more exciting, at least more memorable. The process seemed to have stopped short. I wondered if I had forgotten something. Maybe Lenny forgot to mention part of the secret.

When the rain stopped, and I had caught my breath, I stooped over and picked up the water trapped in my plastic food wrap. At least I could show this off.

I was going to Lenny's house when I messed up though. I slipped on some mud and fell flat on my back. Then the water I'd trapped inside my food wrap fell on top of me. Right away I felt giant's feet pounding inside my head and body. I forced my eyes open and saw that I was being pushed through the trunk of the sycamore tree in our yard again. I could feel my body twisting and growing and spreading through the whole tree, up to the very top limb. I remember the grain of the wood being smooth and consistent, waxy against my skin. I went shooting out the top of the tree and fell to the ground. This time through I knew what was happening. I understood that it must take awhile to learn how water and logging and trees work together.

It was about an hour later when Lenny came walking by. He found me crumpled up on the ground looking like a baby in a body diaper. I was wrapped up tight inside the piece of plastic food wrap. From far away, Lenny said I looked like a puddle of water. Actually, I felt like a puddle of water. I felt limp, soggy and without a place in the world. I was feeling kind of lousy, as if I had messed around with the secret a little too much.

While I was wrapped up on the ground, Lenny stared at me in about three different ways at once. He seemed to be feeling a little sorry for how I looked, but also impressed that I had tried the secret and even more amazed that something had actually happened. Most of all I thought he was a bit embarrassed because he hadn't believed that Uncle Wendell's secret could really be true.

Quick Looks Into The House Of Sky

There are countless numbers of doors into the house of sky. Here are just a few that I have opened, politely, always knocking first. There are many more flashing by all the time. Look up at angles into clouds and you will spot them yourself.

A tornado has a dull, calm eye, as does a hurricane. Imagine being inside this and looking up into the flat, blue sky high above. Maybe there's a bird trapped inside, flying, swirling, easily, quietly. Like drapes flapping in the wind of an abandoned house. This is a vision into the house of sky.

Think of thunder as a language. Not just as something you hear, or count between lightning bursts to determine distance, but as the talk of Rain and Sky Lords. Record these detonations of sacred language, decipher the variety of staccato rips and tearing riffs. Communicate with these high knocks of hammers by synchronizing your breathing, your heart beat. Perhaps you could align yourself with the planets, the moon, the stars and be able to look deep into the quick illuminations of lightning, peering all the way to the center of the sun. Then again, maybe lightning is the slight alteration of a constellation. To understand the language of thunder is to have blind faith and to trust the realm of air. If you can think like this then you have had a vision into the heart of the sky.

Trapped wind. Pick up a sea shell, place it in your hand, next to your ear. Close your eyes and listen for the sound of air. Listen as long as you want. Slowly there will be a pattern. The roll of the ocean, the expansion of the soil, the flow of water, the till of the land, the movement of a bright, red lady bug. This is a beat, a pulse, the very blood at the center of the heart of the house of the sky.

It was probably a trap being set in some way or another.

My brother Spur had already warned me what she would say.

And she did.

Aunt Mar told me, "You're fooling yourself wanting to leave this place. I can promise you that as sure as a rooster breaks the day that all you're going to find is a problem staring at you."

Then Spur said, "Told you that's what she'd say."

Vaguely, for a millisecond, I was feeling smothered, squashed and tiny. I couldn't say I liked Aunt Mar's hopes for me.

It seemed everyone had something to say when I was getting ready to leave Doorall Township and all its narrow, deep valleys and hills. I was planning on leaving the area along with my wife Mulay and our son Roynet. It had been thirty-seven months since the last person had left Doorall, moving along, out into the bigger parts of the world. Scott, Ken and Dean Lucher planked a new 2 x 6 on the Skunk Creek bridge to help us out of the township. Abe Norman unlocked the Doorall Township gate onto

Highway 56. It seemed almost everyone wanted the three of us to make a good head start out of the township.

The oldest man in Doorall, Mr. Lystoder, told me something too. Before we left, he pulled me aside, asked me over to his house, and spoke a piece of what he knew. He told me that he wasn't sure what I was thinking, but that I shouldn't be mistaken. Things didn't grow away from Doorall, they shrunk. He said that as far as he knew, there was no bigger place than right here in Doorall. Something deep in me agreed with what he was saying, as, over time, Mr. Lystoder usually was right. In the case of me moving though, I hoped he was wrong.

There is more to know about Mr. Lystoder and why I feared his opinion though. You see apart from what he had told me, there were his everyday acts and reputation. Mr. Lystoder was bridge partners with Aunt Mar, which may seem small, but really, it's not. The two of them had a certain way of winning the kitty that was strong. People said that the two of them played like they were dealt the hands of God, whereas everyone else seemed to be playing with random, assortments of meager chance. Not many people around Doorall were sure how just being fair could always be so rewarding in a game thought to revolve mostly on luck. People got pretty disjointed about Mr. Lystoder and Aunt Mar. Folks talked about them behind their backs and for that matter, to their faces, from time to time. Like during the monthly Doorall Township Chili Dinners. After a few drinks, Heck Peters would stare Mr. Lystoder in the face and call him a cheater, a two-faced fink.

Mr. Lystoder could always calm the talk about he and Aunt Mar when it got to be real angry though. He was a bit like a hypnotist. He would say things like the two of them thought alike, or had corresponding Zodiac births. Once he said that the reason they won bridge was because they both knew how well a short pause and a deep breath could straighten a person's thoughts. Another time Mr. Lystoder said that he and Aunt Mar could both remember anything that was made up of no more than fifty-two parts divided by four. This comment sounded like Mr. Lystoder was making fun of other people's brains which made everyone a bit uneasy, but then Mr. Lystoder added, in a scientific way that is. People knew science wasn't a joke, so they calmed down. Over and over, Mr. Lystoder had a way of convincing people away from strange ideas. He could put doubts to rest.

Since Aunt Mar and Mr. Lystoder were so close it crossed most people's minds to ask why the two of them weren't married. You see, in addition to being bridge partners, they were also painting partners. Mr. Lystoder explained their union in painting as nothing more than a good mix. He said that Aunt Mar understood color combinations, visualizations and proportions. Whereas he knew how to explain settings and elaborate tones with words. Aunt Mar could match his words to canvas. Together they made art so real it seemed alive. In fact, when a person went to see one of their paintings it was eerie, reality seemed to be paused and confounded. It became clear that what Mr. Lystoder said about he and Aunt Mar was true–they did seem to have a psychic connection. They painted so real that a person would swear the two of them might be in touch with voodoo and fortune telling.

Their paintings were so real, in fact, that when people around Doorall needed to explain the unexplainable, or encountered a stroke of bad luck, they would point to Mr. Lystoder and Aunt Mar as the cause. One example had to do with my cousin Lamar and his wife Jewdell. It started when they were having problems. They both wanted to have a baby girl, but all they had so far were five boys. So one day they decided to have

Mr. Lystoder and Aunt Mar paint them a baby girl resting in Jewdell's arms with the smiling face of Lamar next to her. Three weeks after the painting, Jewdell announced that she was due with a child in March. Low and behold, it turned out to be a girl.

People had fears though. Some people thought that Mr. Lystoder and Aunt Mar went so far as to take life away from things. That they liked to abuse their talents. Some people blamed last year's drought on the two of them. These thoughts were about fifty-fifty in and around Doorall: fifty percent thinking it true and fifty percent thinking it a crock of crap that two old folks could make a drought happen just by thinking and painting. Anyway, the reason the two of them hadn't married was because Aunt Mar said one marriage was enough. She claimed to be married to colors and Mr. Lystoder never said much to that. He wasn't known to argue a point with Aunt Mar.

Reflecting on what Mr. Lystoder had said about things shrinking outside of Doorall Township was not easy for me. I was hoping for things to grow when I left. I was hoping to become a painter as good as Aunt Mar and Mr. Lystoder. I wanted real things to come alive on my canvases too. I wanted to create new worlds for people. I had discovered an idea about painting I wanted to try. My idea had to do with leaving a back door on the painting so that people could walk inside and move around and see life out the front of the painting. Of course, I hadn't told Mr. Lystoder, or Aunt Mar this. I was too scared that they wouldn't take kindly to the idea of another painter in Doorall.

I also had a feeling that the two of them had a mean streak which I didn't want to be part of either. One reason I thought this was because my friend Shane Freed told me he had seen Aunt Mar and Mr. Lystoder make all the cisterns in Doorall dry up in one of their paintings last summer. Shane said he was making a delivery of flowers when he'd looked in their window and overheard Mr. Lystoder mention that they ought to try draining all the cisterns in Doorall. Shane said he saw Aunt Mar paint a dried up cistern, all hollow, gray and empty. Then the two of them laughed hard for a few minutes.

When Shane told me this I noticed he was whispering. He said he was uncomfortable talking about the two of them and wasn't afraid to admit it. He was sure they were like two pieces of black magic and could probably read minds as well. Shane had decided that when Mr. Lystoder and

Aunt Mar painted, they did so because they were mad at people and desired to do damage.

Shane also reminded me, in case I wasn't believing him, that he had played bridge with them and I hadn't. While telling me this his eyes were wide open and he was looking over his shoulder. His energy was nervous and he looked scared, as if he was talking too much. As far as Shane was concerned, the two of them weren't satisfied with reading minds, they liked to turn hopes and smiles into gloom and tears. Mr. Lystoder and Aunt Mar's idea of fun was making other people suffer in horrible ways.

Even though I had a hunch things might get tricky out in the big world away from Doorall, my wife, my son and me pulled out anyway. We drove across the repaired Skunk Creek bridge determined to push ahead and not look back. In our rearview mirror was the rise of dusty gravel and dirt. I felt better just being in the car, wrapped inside the steel frame made me feel capable of what some felt impossible. I thought to myself that I wouldn't be afraid. Even as I thought this though, I began to hope I would never lose the feeling of confidence I had while driving out of Doorall.

After driving in silence for awhile, Mulay said something. She said she had felt strange, driving out of Doorall township and onto Highway 56. Leaving made her feel kind of small and alone, not as happy as she thought it would make her feel. She also said something that made me lose a heartbeat. She said that once we started driving on cement she could hear Mr. Lystoder and Aunt Mar laughing at us. Roynet said he could hear them too and started crying. I figured they were saying this because driving had never been so quiet for any of us before. We only had gravel and dirt roads in Doorall. The smooth, hard cement of the highway was probably just giving us all some extra room for thinking things up.

While I was driving I noticed that the road was stretched out in front of us as far as we could see. I pointed out the long distance of the road in front of us to Mulay and she looked puzzled. We all wondered why the road got so small. Having lived in the tight and narrow hills of Doorall so long, none of us had ever seen out in the distance so far before. We all hoped our car would fit through when we got there.

Roynet said, "I bet we get stuck out there in a tiny dot hole." It hadn't taken long, but here I was in the big world, already hoping that I'd shrink.

After driving awhile longer, Mulay directed me to stop at a rest area. We were all in need of a break. Mulay and I wanted to use the rest room and Roynet wanted to sprint the sidewalk for awhile. When we pulled to a stop there was no one around, and we all got out of the car. When I got to the bathroom I noticed something strange. That the ceilings weren't as tall as the one's back in Doorall. I washed my hands then went back outside. I met Roynet, and raced him back to the car. Mulay came back and suggested that we spend the night at the next rest area we found.

We started driving again. Roynet said it looked to him like the dot hole at the end of the road was getting closer. Mulay asked him if he wanted to sit in the front seat with us. He said yes. I thought Roynet might ask me what would happen after we got to the small dot at the end of the road. He didn't though and I was glad, because I would have had to think about something I didn't want to: that leaving Doorall might not have been the best thing we could have done.

After another hour we pulled into the next rest area. It was dark out and we all felt sore and isolated. We hadn't seen another car, or another person since we left. As I got out of the car and began to stand up, I felt something hit my head.

I couldn't tell what it was. I had to crawl to the rest room, but my hands were too big to grasp the handle. Even Roynet couldn't open the door.

When Roynet and I got back to the car, Mulay was talking out loud about something. We heard her saying that her heart felt punched in. Then Roynet said the same thing. We all sat down in the front seat of the car and because we were scared, we developed awful looks on our faces. Then things got worse. I noticed our car had begun to shrink around us. Mulay was being squashed and Roynet was hitting the dashboard and saying something strange. He was telling Mr. Lystoder and Aunt Mar to quit packing us in with the sky. I hunched down by Roynet and Mulay and heard Mr. Lystoder talking to Aunt Mar. They were doing a painting of the three of us.

I heard Mr. Lystoder telling Aunt Mar how he wanted Mulay to be squeezed between the front windshield and the seat she was in. How he wanted her to be able to peer out and watch the morning sun rise. Then he said that he wanted Roynet leaning against Mulay's shoulder, with only his left eye able to see out, and that he wanted him crying real softly, no louder than a light whimper. He wanted me to face the opposite way, just barely able to see Mulay and Roynet from the corner of my right eye. I felt myself being twisted and couldn't stop the door from caving in on me.

I sat there quiet. Angry. I couldn't understand how we had become no more than a little part of their world, no more than dots of paint. I heard Mr. Lystoder say that he had let us come in through the back door of the painting and he was going to throw away the key. Then I heard he and Aunt Mar chuckling. They both said out loud, "Not bad, not a bad idea at all, but I tell you what, let's not put a back-door on this painting." Then the two of them broke out laughing hysterically.

That night at the rest area was not good. In the morning all three of us realized that we had been tricked for some reason. I figured it was because we were trying to be something extra by leaving Doorall. One thing was for certain, it didn't look like we were going to be able to see what life was like away from Doorall. All we'd become were colors on the front of a painting. All three of us just little brush strokes, immobile, but still painfully capable of thought.

Sometime during the night, while we had all been asleep, or not paying attention, Mr. Lystoder must have suggested to Aunt Mar that she paint a rooster on the hood of our car. The reason I say that is because when I got my senses back, I saw a red rooster staring at me. Somehow that rooster felt like a cold slap in the face. It wouldn't be quiet either, the rooster just kept calling and calling for me to wake up.

Of course since none of us could move, all we could do was watch and listen. When the rooster wasn't making noise, the only other thing we could hear was Aunt Mar washing her paint brushes and Mr. Lystoder talking about us and the rooster.

He said, "I think this is a masterpiece honey, this has to be our best yet. People ought to learn to live at home and not get carried away with big ideas and such." They talked as if they had done us a favor. I hoped it was all a dream. I hoped that my friend Shane wouldn't visit their house and see us hanging up on their wall.

Winter Air — Stone Beating

It is nearly dark, yet it is only three hours after noon. It is stone cold. The sunset is pale and my shadow looks bluish-green on the snow beneath me. The wind blasts into my face and I lean a bit further forward, into the wicked, frigid, tree shattering air.

My shoulders are hunched up, my neck is pulled down so that my chin is resting against my chest. My arms are clinging to the base of my stomach, my body language is all about protection. Yet, I long for these days regardless of the pressure it exerts on my shoulders and neck, these are the good days of the year. The cold ones. The days when it is completely still outside, when life is mostly stationary. In a corner of a field, out of the wind, there are a few squirrels sitting still. There are a few birds clustered in the pines, but they aren't moving, they are waiting.

In this weather I have carried a large, hollow drum with me. I am walking out to a distant hilltop covered with pine trees, I want to be there by sundown.

After ten more minutes of walking I finally start feeling warm. My body has adapted to the cold weather, this warmth stays with me as I reach the hilltop. When I reach the cover of the trees I begin to strike the drum, palm first, through mittens. The noise sounds like it is coming out of a grotto built deep inside a steep mountain cave. I can spot the sound waves of the drum beat and they look like wispy candle light, fluttering with the wind. It is like the sound of muffled gun shot, an empty, earthy crackling. After awhile the sun is completely down and the wind picks up even more. It is blowing so starkly, so stoically, so proudly, so sure of its coldness.

It is the feeling of an old movie, it is gray, yet exciting outside. I continue to pound on the drum. As I breathe, clouds emit from my mouth, which then rise. I think my breath is a thin vapor, a stream of myself, floating upwards. It is a winter air and I am in it, in fact, I am partly creating this air every time I breathe.

I am making the empty, hollow sound over and over and now it is totally dark out and I am still making a beat. It is so cold around me, but the drum beat is like a glowing, radiant, red hot fire. There is a full moon and the land is clean, crystal, crisp, snapping and I stop the drum beat. Then it is silent, but for the wind moving clouds across the moon.

I am just a shadow out here in the light. I move and the snow crunches up under my feet. The snow is so dry it sounds like dry cotton rubbing inside my ear. The wind fights through the pine trees, slightly stalling in the thick, stubborn stand of wood and needles, then finally whips through the trees. Regaining speed, descending, pushing, powerful, down the valley to wail across a lake nearby, creating an eerie, cracking song, a bleating whistle, as it snaps over the ice. A little snow falls off a branch, but I continue to move. I am carrying a beat inside me now.

I watch as loose snow whirls up into the air, then falls gently to form a snowdrift, into what I call the wind's footprint. Again, up above there is the bright white moon, just a frozen light between my beat, my feet and the heat of the house I am walking toward. The wind continues to blow and I breathe and a small cloud forms between myself and the moon. But nothing seems to change, it is still cold, it is still the light of hard stone, of shadows. I recognize that this dull, crisp light is the realm of shadow dirt. Of a place made mostly of careful, cautious motion, eternally, just a shade or two brighter than pitch dark.

I hum to myself the beat of the night in cadence with my steps. I am certain that this is the coldest day we will have this year, that this night is as far away from heat as possible. I keep the rhythm of the drum and the hard silent earth in my mind. Tomorrow it will be a tiny bit warmer and with that everything will begin back toward life again. I think of the beat and hear the muffled earth sing back to me. The pattern I hear is a round, one part of the earth singing what another has just sung.

The Sycamore Throne

They are still out there, people searching and experimenting with the creation of paradise. Individuals trying to create gold. People who reside on the edge of town, or in the "boonies," concocting potions and ideas that will lead to some form of El Dorado and ultimate bliss. Some people are so obvious that they have lost all their belongings, they mutter to themselves about what they know, about how no one understands, how the world has unfairly passed them by. These alchemists are desperate, hoping for golden coins, not golden enlightenment. The type of alchemist I am speaking of looks like you and me. They would never go around in black clothing, or wear gothic jewelry because this would draw too much attention to themselves. One alchemist I know lives in the little-known Iron Mountains of southern Minnesota. He just walks around, fishing mostly, but in the evening he is busy tossing together the ingredients of the world, panning for gold with the wind, water, sun and earth, examining open trees and looking for treasure.

These modern-day alchemists operate on a highly imaginative, self-conscious level. They are master mind readers, very close to being Shamans. They are positive people. An evil alchemist won't disclose anything to you, and is highly critical at all times. They thrive on back stabbing and enjoy catching you making mistakes. A positive alchemist is making magic all the time. You may think you are speaking with one if they talk to you about their imaginary friends, or if they have a lot of dirt on their hands since they are frequently gardening, or working with wood. Another clue may be that they help you without you even knowing, offering advice you end up needing a few days later.

It is the angry, the bad and the mistreated you must stay away from. If a person makes you feel depressed regularly get away! If someone doesn't allow you full expression, makes you feel a stranger in your own home, expel them instantly! These people are working a potion of deceit and would rather you fail than succeed.

Especially telling are people that have products, or livelihoods which are supposedly based on excitement, and individuality, but yet when you meet them, or talk with them they are boring, tired, or ungiving. Recognize this, for they are only pretending to be alchemists. Alchemists are naturally capable of invention, they don't need to try. They are gold, and from this basic and healthy quality they expand.

It has been said that the key to gold and enlightenment is visualization. Many mystics hold their hands above their heads when preparing to engage in creation. I, mind you, am not an alchemist. I simply try to find them and make visits. If you were to ask me how to get to an alchemist, this would be my answer. I would ask you to follow me back to the Sycamore Throne, where I was lucky enough to find my first. Close your eyes and listen. Let me take you there.

Journey to the Sycamore Throne. As you are walking one morning, a trail is initiated by chance. Tacked to the trunk of a large and craggly oak tree is a small wooden box turned sideways, angled away from what is often a stiff north wind. Inside the box is a tattered and brownish, yet neatly handwritten card which reads: *Paron Beet — Tree Dweller, Branch Hermit, Leaf Speaker, Root*

Eater, Green Spirit. The box and the card are an announcement of sorts for a hidden, almost completely lost area. A place, some say, which is older than the first human grunt.

The path leading up to this box and the announcement is very singular. It is so vague that it can only be perceived with a little luck, and in just the right light, at just the right angle. It is an astute observer that will notice the lightly tread upon entrance, the tiny, bent blades of grass and the slightest of matted down moss. Fine attention will inform a person that the path is mildly held apart along its way by the pull of spider webs. Taut strands of silk tugging maybe an inch to the side. It seems unlikely that the trail has been used with much regularity.

On the summer solstice the trail is more likely to give itself away. For during the half minute to a minute when the sun probes the furthest into the land, the trail is clearly shone. The lightly matted grass and moss shine. There is also a small, although brief, reflection given off by the nail on the box, which is holding the card in place. The bright, mirrorlike reflection could easily catch someone's eye. It has just caught your's.

The path slopes down quickly once you leave the card. It nearly drops you thirty feet. It isn't clear at first what Paron Beet could be, but you assume it is a person, and that the path will take you to him. The forest is so dense that it is impossible to avoid feeling the trail is not leading to something Holy, or Sacred. You wonder if perhaps Paron Beet is a Saint of some sort. This only seems exaggerated in theory. The lush and fruitful forest, so moist, leathery and charming evokes a real and deep feeling. There is a smell like fermented cloves, vanilla and cinnamon in the air, all of which further convince that Paron Beet and this forest are blessed.

A pilgrimage, then, to see this Paron Beet and these trees is in order. To journey into this land, which feels thoroughly innocent and virginal, may allow you to understand, without speaking a word, such things as intuition, creation, holiness and magnitude. You imagine, already, that the only way you could ever describe this place would be to use mostly body language. You could say heroic, but then you'd have to look out at the sky while holding your arms out as wide as possible, behaving bold and brave.

The trail bends around a corner shortly after you feel plunged into the forest, after you feel dropped into the massive trunks of wood. The physical environment is subdued and barely moving. There are no tricks being played on your expectations. Looking back up to try and spot the sky, the branches and gnarled sycamore tree trunks get in your way. You struggle to view a clear path back out but you can't. You feel inside a spider's web. The presence of the trees is not a lie. They are obvious and overwhelming.

Continuing further, the trunks of the trees take on more mass. Sunlight is filtering slowly past the branches and the leaves. You start to count the variety of light, but you stop at thirty-three shades of green. The journey is well under way. Your memories are long past useful, so much so that even hallucination will do you no good.

And their is a noise. The heartwood of the trees, of the oaks, the pines, and of course the sycamores, even some cottonwoods and even strange redwoods, are all soaking up water so loudly they slurp. On the trunk of one of the trees is another box, with a handle this time, and a door to open. On the inside is a list and a small note. There is also a slim candle burning with a cold, smokeless flame which reminds you of your home in the city, where, when the electricity goes out, everything suddenly becomes so calm and quiet, lit by candlelight.

The wax burns clean and the tidy appearance of the box makes you believe that someone probably has to change the candles regularly. It makes no sense, but then you understand that maybe this is the point, that maybe your mind is being reinvented and stretched. You look again and the box just seems too small to contain a candle and a flame and a note, but it does. Again you confirm, there is no indication on the inside of the box, above the flame, or anywhere, of heat. Not even a bit of a tarnish, or for that matter, any smoke. Just a cool flame, like glacier ice, glowing blue-green, eternally. The note inside contains the following:

My first ten commandments.
I mostly got new ones now though.

1.Don't you worry none much son, we'll do the big gun on 'em if they come for you.
2.Ain't nothing made of what it seems.
3.Jumping off high places is safe.
4.Talking fake languages throws your opponent off. Moral: babble be gooden.
5.Rabbits got stone feet.
6.Hunting ain't what it used to be.
7.Probably won't rain and if it does wouldn't much matter none.
8.Sticks and stones break your bones, but they can also be made into better bones.

9. We got bad news Paron, your Brother was a Hoot Owl.

10. Your great Nephew, Flick, he discovered Overalls.

Note: These few things are all I really learned about life from my folks, who being of sound mind and body, did have more to offer me, but didn't. Nothing new in that I suppose, you get in a pattern and things don't change. You begin to wear the color blue and after enough people see you wearing blue you feel a certain comfort and attachment to stay this way. My Pop, not being mean or nothing, started saying the same dumb things to me and never stopped. I learned to put together unusual thoughts and cope as well as I could with normal people and ideas. But, all of this is boring, you can probably tell already that things are off plumb a couple of bubbles with me, so move ahead, on into the trees, I'll be waiting. — Paron Beet.

Moving further down the path, the trunks of the trees continue to increase in girth until it is doubtful that you will be able to slip between them much longer. All available space for movement is being taken up by the broadness of the trees. It is still easy to breathe though, in fact, the air is amazingly comfortable, smooth, like breathing air made of polished silk.

The exaggeration of the silence around you is hard to miss. It is so quiet. Not the kind of silence you sometimes get early during the morning, in a city, when no cars are moving. It is much quieter. The silence around you has no comparison at all. What you can see of the silence is that it is made of a variety of light which has been bouncing around, underneath the cover of the green leaves for maybe six, or seven years. Light, once trapped in this forest, lives until it creeps and finally forms into scuff marks, or subdued glows, on the bark of trees. Sometimes, a little of the light continues to float in the air and gets inside of bugs which flicker from time to time and keep the light around even longer. Yet, for the most part, the light bounces and fades and slowly falls and hushes the surroundings.

Moisture obviously gets trapped as well. Water has been dripping down from the tops of the trees since before the Civil War. You spot a drop of water now, that you would like to touch, but can't quite reach. It must be so well filtered, having dripped down through the forest for so long. This forest seems a higher form of life where nature's objects are able to completely fulfill their destinies. It isn't long before you run upon another box which contains yet another message from Paron Beet.

Warning: Down low in the valley is where my trees are growing. I only call them mine, they aren't really, they belong to the world. These trees are growing away from commotion. I raced the wind to the bottom of this valley decades upon decades ago. I am still waiting to find out if I found the bottom. In the meantime I've learned to talk tree, heron and water. I now know the faith of nature and the belief of rain drops–Paron Beet.

After reading this note, the only problem with continuing down the path is that the tree trunks are now growing into each other and there is no obvious way to keep walking. You'll have to figure out a new way to travel. The presence of the trees is becoming so strong, however, that you eliminate a few ideas which might ordinarily make sense, such as sawing, cutting, hitting, or hacking. You bend over, lay your head on a massive conglomeration of tree trunks and listen. The slurp of moisture is still going up. Behind the pull of the water is a slight murmur, which you decide must be the trees creating fresh air. It crosses your mind that, since the human body is mostly made of water, if you stay too close to one of these trees, you might be mistaken as a small watering hole for an oak tree, or large redwood. Perhaps this wouldn't be such a bad occurrence though. It would be more of a compliment really, able to blend in so well with your surroundings. The only problem is that you want to move down and the water seems to be going up. You watch your step, in fear of finding yourself rising through the trunk of a thirsty tree.

You look around to find a clue. How will you be able to continue moving down? For a moment you wonder, where are the birds? The squirrels? After another moment you notice woodpeckers, slipping up and down the tree trunks. Is there a clue in the movement of these birds? Sitting down on a tree branch, looking around, you suddenly spot another small box, about fifty feet away. The box is tacked to the side of a tree trunk, well below the tree you are sitting on now. Your only choice to reach this next box will be to jump and climb, like a squirrel. Moving gently from limb to limb, sliding and pulling, moving further down into the realm of Paron Beet.

So, you do so, easing your way down the twisted tree trunk, over to the next tree, gripping along as best as you can. Getting closer to the next note you notice that there is a cluster of fire flies moving back and forth, all around the outside of the box. Moving further, you reach the next note. You hope there will be a clue which will let you know how to get to the bottom of these trees, it is unbelievable the forest floor is still no where in sight. Next to the box is a wide and gruff burl which is a ledge for you to stand on. Standing up straight, you peer into this next box of Paron's and realize that there is no way out now. You must trust what is below you, as the way back up is far too steep to climb. Opening the box you see that, like before, there is a candle slowly, mysteriously burning.

Me be no mutt. Me be honest abe. I got grit. The weather outside is turning. NOW LISTEN. Most games are win or lose. Most loses are shame or hate. Most wins go away. From here, you can fly, at least it will seem so. I do. Lay down and follow the smooth feel of the trunks of the sycamore trees around you now. Slip and slide if you're a smart one. But first, since this will be your last stop before you see the Sycamore Throne, close your eyes. I can't be easily seen: I am chalk dust, or the center of a blue heron's eye. A long time ago sycamore trees were the markers of God and the anchors for the stars, they were where constellations lived during the day time. They were hollow chasms, with such girth that their tops were cloud formations. Their trunks are white and chapped from so much movement through clouds. Squint your eyes, look up, spot the trees up high. The trunks of the great sycamore trees are straws, letting the earth breathe, letting the light and heat of the earth's core spew forth into the universe. At least this is what I believe and you may too when you reach the base of this land. Below you the great sycamore trees are resting, all downy, white, and full of great messenger birds. Jump, slide, trust me. Don't believe your mind. Ask yourself if you can fall off the earth once you're already on it–Paron Beet.

So, you close your eyes and just let go, and because there really is no empty space left between the tree trunks, you are sliding down the trunk of a tree. Then you feel a wind. Then flight. Then there is speed, but you keep your eyes closed because you know this is a time that you have to have faith and trust. There is something guiding you slowly, rocking you like a baby, down.

Finally, you stop and are facing head up, staring at the flight of hundreds of blue herons. The birds are nesting in the tips of a colossal ring of smooth and polished sycamore trees. The ground crunches beneath you, you're on top of a large pile of bones and leaves which have fallen from the herons. When you stand up, the birds all reposition themselves inside their nests, watching. There is a mass guttural squawking. A hand places itself on your arm and you go without looking. You can't see anyone, yet you know your eyes are open and you know it is Paron Beet. It is the same sort of strong, invisible presence that he has had since you first chanced a view of the path leading down to this ring of trees.

Eagles have begun to cluster in trees behind the herons and there is a river running nearby where the trout have all stopped in place and are watching. The word intruder might as well be amplified, screamed for all living organisms to hear. Paron, you assume, pats, then pounds on the trunk of one of the trees and a noise, the strength of a hollow cave turned drum, beats and moves the air with gusto. The echoes of the slap on the tree fade until the realm of the Sycamore Throne is still. Then, after a moment of silence, Paron starts to clap and all the birds respond by taking off in circular flight.

Leaves fall off the bushes near the ground, as the great blue herons, the eagles and even great horned owls flap their wings. He must be a shaman, or an alchemist, this Paron Beet. He seems to speak for silence, for herons and for trees. He glances quickly at you, too quick to really get a good look. He looks through you, and then has moved to a mound of white bark chips and dirt. He places a crown of sycamore limbs, seeds and feathers on his head. He is now in the middle of the ring of sycamore tree trunks and you notice that all the heron's have landed and seem intent on listening. You think again, Jesus was a tree and so maybe this is the Garden of Eden. You wonder mathematically, is flight equal to air moved by the speed of sight? Questions fling through your mind and then the drop of water you remember from earlier, when you were up higher in the forest, hits you across the forehead, drips down your face, moving you out of your trance.

While Paron Beet is sitting in the throne he tells you to move to the river nearby and stand in the water. The water is so cold. You begin to shake. Your neck quivers. You stand still in the stream and all you can feel is your feet. You look back to the throne but Paron is gone. You stay in the river though, waiting, trusting that you will receive more directions when needed.

For now, standing in the river, concentrating, you can't fathom how many years it may be before you leave this realm of Paron Beet's. Your mind is alive and filled with rhyme: eternal water, green light, reverent trees and strong, bold flight.

This thing you are doing now, standing, listening, concentrating, putting down roots in the river bed by the Sycamore Throne is so much better than your normally hectic life on the earth's surface. You are wonderfully involved in an hermetic union with the natural patterns of the world.

Sometimes, in the following years, you spot Paron Beet standing in the river with you, or you spot him evaporating upwards, to replace the candles you passed on your way down. Such circular movement is comforting. Paron can materialize and create himself both physically and mentally. He is just as much a ghost as a deer, just as much a thought as a common human.

Finally, during this venture into the river, after what you assume has been many years, Paron says you can leave. So you walk out of the river and build your own box, and write your own note to leave behind, along the trail. *"Trust beyond the quest for gold."*

You roll this short sentence up and place it in the forest, so that on some summer solstice day, or some cold, clear autumn night, it will illuminate and encourage others to take a journey of their own.

Buddha Boulder Builder

If you look along the river bank right now you will spot Krick, a man stuck looking at the continual flow of water going past him. He is probably studying the molecular formation of atoms and how they maintain themselves in the cold water current. He is sitting face first into the sun. He refuses to cover, or close his eyes and by now it would be a fair assumption that he's probably gone blind. No one you ask about Krick's behavior seems too concerned though, they just say, "That's the way Krick is." Or, "If he wasn't a character," as Ms. Minestorm always says, "you probably wouldn't even be here anyway."

It used to be that the town of Peru was a bore. People moped about the streets thinking they deserved nothing and that their small town should just blow away forever. After all, when you aren't smooth and slick, when you don't have a local high school and even your town cemetery has been lost, you pretty much have nothing. Nelson Morgan's kids cried so much about not having cable television that he sent them off to live with his brother in Rochester. Even the fast food mart gas station, on the outside of town, had gone out of business the winter before Krick had done his first thing.

Two years ago, Krick was turned from fool to hero in about three days. Before Krick's transformation, things were so bad, that even the last second growth tree in Peru had been cut down behind Aldo Hash's house. The city council had decided that cutting down this last tree would be a symbolic way of ushering in a new way of doing things. The old council members seemed to believe that this act might bring a sort of reincarnation to everything. Aldo was the proponent of the whole thing. He declared that felling the tree would bring about new growth for the town.

It was something to behold when the last second growth tree was cut down. What it meant was that some of the older people had started giving up on their old fashioned beliefs. It was sort of like the old folks had to admit that maybe they had things wrong about how to maintain a town. Maybe what they had thought of as rational and proper thinking, in the past, wasn't that at all.

Standing on the back porch of his house, staring at the fallen tree, Aldo Hash was certain his idea would work. Aldo told his friend, Abraham Sleuth, that hope can sometimes be all you need. Abraham told Aldo that sometimes hope is all you have. Aldo told Abraham, we're bound to be better before too long, all we need is a little twist of fate. Abraham slapped the hard oak paneling next to them both and nodded his head. Then they both tossed down full jiggers of scotch.

Krick had no idea that Aldo and Abraham were having this conversation. Earlier that day he decided that since he was bound to stay in Peru forever, and since no one really cared much for the town anyway, he'd make the place more than just a memory. He was going to make it right again. Put it back on the map, literally. Peru had been left off the state road atlas now for close to thirteen years. Krick had no special skills, but he didn't think that it would take much to call this place his home. Just a little positive identity was all the place needed. He may not have made it to some far away college, but he knew a good thing apart from a thing called boring, so he went to work.

He walked directly in front of the Ice Cube Cafe, right in front of where all the old folks looked out when eating their afternoon dinners, and started spinning around and around, not real fast, just around. He had told himself that when he first started twirling he'd get past the dizzy part within the first couple minutes, then see what happened next.

Abraham Sleuth, owner of the Ice Cube Cafe, looked out the window and didn't say anything at first. Then, after Krick had been twisting for a good five minutes said, "What in tarnation is that kid doing?"

"Looking for attention, what do you think?" said Melton Throwball.

All the patrons of the Ice Cube Cafe turned their backs after a bit and only occasionally looked over to see if Krick was still at it. Wand Bicker, after everyone had been quiet for a bit about Krick, added, "What this town really needs is more people like Krick." Everyone paused for a second, holding dead pan faces, then broke out laughing, real hard, including Wand.

Aldo Hash on the other hand thought that it was kind of beyond coincidence that Krick had shown up, out of the blue, shortly after they had finished cutting down the last old tree. Between he and Abraham, they knew it was going to take something to make a change around Peru and Krick might as well be it. He kept his mouth shut about this though.

In the meantime, behind the old folk's backs, you might say, Krick attracted a crowd of younger people and a strange dog that had stretched out on the ground to watch. It turned out, as the sun lowered in the sky, that Krick had fastened colored mirrors all over himself and so, as he spun, colors bounced off of everyone's faces and bodies. It was sort of hypnotic, everyone simply stood without thinking, just looking.

After everyone had gawked and talked and finally walked away for the evening, Krick kept spinning and the new dog in town kept watching. Right before he locked the cafe doors, Abraham Sleuth looked out his window and had to admit, that as weird as Krick was behaving, he was impressed. He hadn't seen anything being done, for even a short period of time, in Peru for years now, even if it did seem like something crazy.

Early the next morning, way before the sun was above the horizon, Flu Bats, who dropped off the bundle of morning papers for Peru, saw what looked like a tall tower, kind of like a cross between a big tree and a lightning rod, spinning in separate sections in front of the Ice Cube Cafe. Although an overcast and windless morning, different pieces and flashes of color were moving at different speeds, in different directions, making reflections all across town. Flu saw, off to the side, sitting down on the curb Krick and a dog both watching, kind of gawking really, at the spinning colors. Flu dropped the bundle of newspapers into the paper machine, looked a bit longer, glanced at Krick, got no response, then drove on his way.

Later that day, when he told his friend Freddy Wilson what he had seen, he added, "It's kind of strange really, that town ain't never stuck in my head before and I've been going through every morning for twenty-three years now. I mean afterall, Peru, in my mind, that's the sorriest place I know."

That same morning, after Flu had dropped off the papers, Abraham Sleuth came by to open the cafe. When he saw Krick still around, sitting on the curb with the dog, he was shocked. He was even more surprised to see the tower of colors. He realized that Krick had done something all right, he had made a piece of moving art. "It sure is different," he whispered to himself.

Krick walked right through the mirrors, shook Abraham's hand and said, "We need to make a town out of this place, Abe, and looks like I'm going to be the one to do it. It may take a goodly while, but I'll put us back on the map."

"Well, I'll be damned son, what have you done there?" asked Abraham.

"Brought you a dog and made us a miracle," said Krick. "What do you think?"

Abraham said it again, but louder, "It sure is different." On the inside he was thinking, "By God, I think this kid may just be the answer."

The dog saddled up next to the front door of the Ice Cube Cafe and Krick walked off. Abraham looked at the dog, grinned, and went inside. He brought back out some scraps of food. Abraham had never really thought about it, but once the dog was at his cafe's doorstep he was mighty happy to have a companion that he could feed and look forward to greeting in the morning.

After Krick made the spinning tower of colored mirrors, the town started rebounding. The primary function of what Krick had done wasn't really the spinning colors, but that, now, the town had something to boast about. So when everyone saw Krick sitting down along the river this year, facing the burning sun, people were interested, and hopeful.

"Could he do it again?" everyone asked, "Could he invent another miracle?"

Gathered inside the Ice Cube Cafe people had already begun to speak of big things. Abraham Sleuth said he would expect nothing short of a bronze statue.

Wilt Grunt said, "A nice big fire would be all right with me."

Aldo Hash said he thought Krick might just plant a tree. Folks looked at him funny, but not Abraham, he said, "No dumber than pinching someone to grow an inch. Heck, no stranger than the spinning colored tower we already got."

Closer to where Krick was sitting in the sun, roasting out his eyes, were some younger kids. One was Cap Lunar, eighteen and thinking about moving out of town. Cap said, "Betcha he'll become a wooden match. I want this place to light up like a firecracker or something, that's what I'm hoping for."

His girlfriend, Coley Mustgame, added, "He's gonna make us a dot on the map again, I betcha." Just thinking about Krick and what might happen made Cap and Coley start spinning around in each others arms. Cap shouted out, "They're even gonna know us in Walla Walla!"

When Cap's little brother, Tonka, heard Cap mention Walla Walla, he said, "Who cares about them. We got better, we got Krick!"

Krick was blazing bright by the end of his second day in the sun. His eye sockets were hollows of boiling lava. The moisture of his body was leaking out his ears, mouth and nose. During the middle of the second night, he turned into a noise, like the high pitched squeak of dry finger nails across a chalk board. The squeak went on for a good five to ten minutes during the dark of the night.

The following morning, Krick was done with his creation. He'd made what would be one of many more. Where he had been sitting, now rested a large white boulder, about eight feet wide and ten feet tall. Smooth and looking like whipped cream, or a really big, giant puffball mushroom. All around the boulder, daffodils were blooming. Krick stood to the side, took a look and liked what he saw. "Kind of like a seed," he thought. He had to admit, he rather liked the idea of giving birth to boulders. He knew right then that his goal would be to make one large, humungous boulder that would stand like a chapel for the town of Peru.

For now though, as Tonka, Coley and Cap marveled at Krick's new boulder, they and many others wondered how it was possible. "No way! He made a rock?" Tonka blurted out.

"Yeah, he made it. Wasn't here before. Look how soft and calm and smooth it is. It's like those rock gardens they have in Japan or something," said Cap.

"A zen garden you mean," corrected Coley.

"He's really gonna get things going around here, that's all I can say. We'll probably even have tourists this summer," said Cap.

"Maybe we could talk Dad into opening up the old ice cream place," chimed in Tonka.

The three of them spotted Krick to the side, sitting and looking kind of worn out. Coley said something that had not been heard for decades in Peru, just loud enough for Krick to hear, "Ever since we got Krick, we live in the best place in the world." Krick smiled and made no notice of himself after this comment. Coley, Cap and Tonka were all happy to suddenly live in such an important place.

Krick was glad he was making things right and kept at it. He remained busy through the spring and the summer making new boulders about every five to six days. This kept up until the end of summer when Krick stopped for the year, after making a rather large, white boulder the size of six, tightly stacked, straw bales. He thought to himself, "I'll try once more to improve on this, but not for awhile. Maybe next year sometime."

This last boulder of the year, the size of six straw bales, was smooth and round and glowed deep inside when you watched it at night. "The earth and the universe are kind of symbolized by it, I think," said Cap, who suddenly wanted to go to college and major in eastern philosophy and religion.

News was getting around too. Flu Bats was taking a mouthful of gossip with him each morning. He was telling everyone that something was happening in Peru. Tourists started coming by. You could tell they were tourists because now, when strangers came to town they no longer raced through, but instead, they stopped and went slow, following the laws and all that stuff, even talking to the locals and wanting to hear about the miracles going on.

The way the tourists started arriving, agreed the old folks at the Ice Cube Cafe, was most uncommon and exciting. It was like their lives had once again become important and interesting.

Abraham Sleuth was as happy as anyone in the new twist of fate. He said, "I never thought we'd get back on track. I just always assumed that I'd be dying alone, serving two-egg omelets to shadows. Now I think I may even create some new recipes for my menu. I feel like a new man, I want to makes some changes!"

The town began to transform and people liked it. It was something full of pride to go to the state fair and put up the sign declaring you were from Peru and have people from all over the state ask you questions. It was only a few years since Krick had made the tower of colored mirrors, but since then, the citizens of Peru, even the young people, were choosing to stay in town, or to at least return when they graduated from a college. Some of the kids wanted to go to school just so they could learn how to make Peru a better place.

Well, it was about two years after the first boulder had been made when Krick finally gave birth to what he described as the, "one and only, holy boulder." This perfect boulder was born after four extremely cold, but partly sunny days in the early spring. As he stood back at twilight on the fourth day, he knew his challenge to hatch one purely, holy item was complete. Where he had been, now rested a smooth, clear, glowing boulder, standing exactly thirty-six feet high and thirty-six feet wide. This boulder was nestled amongst a cluster of white barked, sycamore tree branches, and most remarkably, the boulder was hovering about six inches off the ground. The boulder sat at a slightly odd angle and seemed to have a glow which followed a person wherever they moved, like the eyes of a great painting. Krick reached out immediately to touch the boulder and it was soft and hard all at once. It was firm, yet forgiving. It was more alive than anything he had ever felt. He thought to himself that it was sort of like an inland coral reef, the first of its kind. He was even able to hear water circulating inside the boulder. After his brief examination, Krick calmly walked away and told himself he would never hatch boulders again, and he didn't.

The most recent news from Peru is that a group of monks, from a region in Asia that Krick had never heard of, wanted to know about Peru and this man who had been able to give birth to earth, water, fire and air.

When the group of monks showed up, an aged and smallish looking gentleman, the leader of the group, asked to be led to the final boulder Krick had created. The small man seemed to recognize that this boulder was outstanding and stopped all his talking, sat, and chanted briefly. He then approached the boulder, where he rested both his hands and forehead directly on the stone for nearly an hour. The others in his group had kneeled and waited. The leader explained to everyone how he believed this boulder was the essence of the spirit of the Buddha and was honored to have found his way here. The local spirit was strong in this area he said.

It was nearly fourteen hours before the group of monks left, but when Krick learned what they had said, he replied, "I don't know for certain what a Buddha is, or even what a monk is for that matter, but I do know that the rock they liked was the best one I ever made. That's why I stopped, I couldn't do any better."

Since the departure of the monks from Peru, unlike almost any other small town in the United States, Krick and the people of Peru have taken on a new sense of pride. The young people, if they do leave town, want to return and grow up in Peru. They want to be hometown lawyers and doctors, farmers and writers, some even want to become sculptors. Everyone feels Peru is an important place to live.

People like Abraham continue to talk about the tower of mirrors which is still spinning. Tonka, now going on twenty-two, fresh out of college, has been going by Krick's house recently, asking how he might be able to start making rocks. He told Krick that he'd taken a class on Buddhism in college, and how he wanted to live in Peru, practicing life in small doses, yet, by doing so, becoming complete. He didn't want to do anything fancy, he just wanted to be involved in basic things, fully and boldly.

As for Krick, he doesn't make rocks anymore, he just goes around town, not like a hero really, just eating and working and sleeping like anyone else, except for the constant demand on him to do weekly interviews on the television and radio. Now, because of the actions of just one person, the town seems blessed and worth keeping track of. Krick is always quick to point out that, "It could happen anywhere."

Ghost Town

(Author's Note: This story is not mine. I found this mysterious tale stuck inside an old bottle, alongside a trout stream, out in the western half of the northern Great Plains. I read the note with great fascination. I could recognize the foundations of many old and tattered buildings nearby as I read. Inside the bottle was the name Decker Tab, and a short note saying "Written On His Behalf." If anyone knows this man, perhaps they could provide further clues to the contents of this message).

The sun and the wind and movement of the earth struck out on their own one day, ignoring seasons and patterns and the equinox. Doing so, a small northern prairie town was slowly removed by the hottest, brightest days yet known on earth.

Brightest and heat took over the town of Fargas Union by surprise. The people hadn't done anything to try and change, in fact, their lives were remarkably average and plain. Fargas Union consisted of the usual collection of folks. Some who claimed to see spaceships, some who worked in schools, some who ran auto shops, there was a little bit of everything to go around. But on the day of the summer solstice, the healthy variety of activity in the town began to cease and a monoculture invited itself in. The sun began to take over and the shadows were destined to be melted into tar. It was a summer that would not halt.

People didn't really pay much notice to the shift in daylight until the beginning of August when the days just kept getting longer and longer, even though the summer solstice had come and gone. Decker Tab, who was the Mayor back then, told his neighbor Neil Flinch, the newspaper editor, that it would be best if the paper didn't start making the increase in sunlight an issue just yet. He didn't want the people in other cities wondering what was wrong with this town. Decker thought it best to keep the story local and take care of the problem from within.

Neil Flinch told Decker, "I don't see that it makes much difference. Who'd believe us, facts or no facts? We could probably share the story, and still no one would ever notice, but yes Decker, I'll keep the story shut for now."

"Good, it's the principal, Neil. We have to work alone as long as we can. The news must not get out." Decker Tab then added, "It's a matter of pride, Neil, pride."

People only got nervous though, not telling anyone else. By the time November rolled around, not only were the days still getting slowly and slowly longer, but brighter as well. This increase in daytime wasn't cooperating with the sleeping habits of people either. Nearly everyone in town had been raised to work hard all day and sleep all night. Now, it was rare if a person was able to grab much more than a couple hours of rest. In December it was worse, some people were getting maybe half an hour of sleep a day. By the end of February, night time had already disappeared for over a week and even the shadows were worn out.

Belinda Conray, a small, third grade girl, squealed one morning when she watched her shadow creep away from her and hide under a rock. It was weird, everyone agreed, not to have a shadow anymore. Yet, Decker Tab, who had not slept for nearly a month and a half, still wouldn't talk to the neighboring towns. Now it wasn't because he was too embarrassed to tell the truth, but because he was so sunburnt and in such a manic state of mind that he doubted anyone would pay any attention to him.

By the middle of April people were just plumb, plain crazy. People were shocked, stunned and couldn't believe the sun was really acting this way. Everyone felt cheated, like they were unimportant, like they had been taken advantage of in some way. That the laws of nature had turned on them seemed a cruel act of betrayal. Folks were out of their minds, trading fingerprints, making fun of their friends, imitating cows, even setting a place at their dinner table for the sun, as though the burning orb were some sort of new member of the family. Pretenses of civility had long since disappeared.

Ms. Viola Green had become so upset with the sun that she had looked right at it, with her hands on her hips, and given it a piece of her mind. She yelled so long that when she looked away from the sun her eyes just caught on fire. Nothing, not even crying heavy sobs, could save her eyesight. All through her wailing and screaming and cussing, the sun just kept on shining. Although, perhaps as a small consolation for Ms. Green, the sun did stop being something she could see.

Nate Copple hid no feelings about how he felt towards the sun either. "Heck, there's nothing proper in this light, nothing. Even my sweat glands are all used up. If not for spit, I'd have no way to cool down."

The sun was playing all sorts of jokes on the people. Laurel Smitterson, who was seven years old, told her mother she wanted to be a sun when she grew up. Gene Rifflebutts came up with an idea to reinvent the sunflower. People had begun to talk of nighttime and shadows as if they were things they used to have to walk twenty miles to see, back in the old days, when they all lived in log cabins without running water.

It was about one year later, a sort of full cycle in sunlight, when things began to change, when the first person dried up completely and shrunk into a small piece of parched skin. Decker Tab had been called over to see the dried up person and he had suggested that they roll the individual up and stick him in a bottle for safe keeping. Neil Flinch added that they should tie the bottle to a rope and let him float in the river.

Over the next few weeks more than half the women, men and children had been rolled up and stuck into bottles, jars and boxes, then labeled as to who they were. Some of the people still alive, like Morgan Bullwhip, Pat Borns, Julius Eden and Gene Rifflebutts, thought that the remaining people in town ought to tie themselves to rocks and just jump into what little water still remained beneath the waterfall in town. Decker Tab was absolutely against this though. He suggested that everyone pitch together and fight like men, maybe make a giant quilt to hang over the town, do anything for shade. Gene Rifflebutts told Decker there was no point, that either decision was suicide.

Morgan Bullwhip added, "Hell, only a fool would want to stay alive around here anymore anyway Decker."

Everyone looked at Decker Tab for a response, but he had quickly dried up and turned to parchment as well. "Seal him up," said Gene Rifflebutts.

"Well, I'd say old Decker was a good Mayor. This sun thing would drive any Mayor crazy," spoke Julius Eden as he stuffed Decker inside an old bottle, then tossed him into the river.

After sealing up Decker, the rest of the people in town, all thirty of them, tied rocks to their arms, legs, necks and stomachs, wrote their names on pieces of paper, stuck their name tags in their mouths and jumped into the pool of water beneath the waterfall. It was at that moment, when everyone was either under the water, or stuck inside jars and boxes that the ghost town began. Out of the blue, exempt from history, the whole town had disappeared without a trace. Taken away by a trick of the sun.

Slow Rain Birth

Nighttime sky water falls without any warning and one morning, spring is here. People wake up in the morning and think, it's raining. Think, the world has been cleansed over night. Think, I could lay in bed all day with a rain like this, so nice and slow and foggy. The ground says, mud. The earth says, seeds come to life. The hidden, underground crevices say, fill up my aquifer and find ways to stick around through the next drought. The crickets emerge from the holes in the earth and will have to wait for the land to dry out to return again. The earthworms do the same, but slower. The stars drop another spring hatch of lightning bugs for the summer. The frogs clap and begin to chirp. The grass goes into it's artist-like green movement. All of this just happens. So that when you awake and stand up and look outside, things are well under way. The sky has mixed with fire, and with air and with earth and with water. Everything is involved in creation.

Grant Wood: Land, Sky and Stone

Let's build a massive stone barn. Let's start a regional art school. Let's wear overalls. Let's be true to where we live. Let's take a look at Grant Wood.

Focus on stone, earth, weather and sky. It is hard to venture into Grant Wood country without a set of paints in your grasp because of the rich feel of the land. The steep decline into Stone City is a combination of prairie, hill, mountain, and valley. The rolling hills and stone buildings are a painting, but they are also fully alive. They stand for a region that was and is still strong of natural character.

In painting, Grant Wood seemed more intent upon the land than the sky, yet his impressions pay utmost attention to color which is a revelation of the weather and sky. The harvested and well bundled sort of landscapes done by Wood are extraordinary. The use of pattern in a Grant Wood landscape makes you able to taste and smell an area like Stone City. Everything looks edible and thus becomes the bread of the place. The edibility in a sense, makes the painting come alive. This wish to eat is appropriate since most refer to the Midwest as either the corn belt, or bread basket. Grant Wood did seem charmed by the air, however, and his painting, Spring Turning, has the clouds mirroring the plots of land. His feel for the region, his passion for the living, growing, heart of the land and for the growth of trees and values of people on the farms is captured both in portraits and in landscapes.

It is easy to say his work is bubbly, or orderly, or dreamy, or fake, but wait, look at a Grant Wood painting again. Go where the Iowa land rests. Then look a little longer and you will see for yourself that the paintings aren't imagined, aren't jokes. Perhaps take the word of a native Iowan, these paintings and portraits are not trivial.

A major compliment for any artist ought to be that of a good local story teller. If an artist can depict daily life and details in such a way as to capture the heroic deeds of the silent and simple lives of people, the hopes and dreams of the future and efforts of work, the growth of trees, the spread of wild grasses, the patterns of wildlife, the glow of the sun and the moon, then the artwork becomes honorable and revealing. Grant Wood's painting is cherished for just these reasons. He knew, appreciated and valued where he lived. At the same time, Iowa created Grant Wood, presenting itself with honesty for his palette.

As Grant Wood said, "I realized that everything I had learned I did so while milking a cow in Iowa." Full understanding starts with a couple things. It comes from a strong sense-of-place, as well as from what cares about you the most.

Weather Report

A high and mighty wind is blowing. When I look up into the sky I see small merganser ducks doing backflips against the air current. There are black crows resting in the tips of high cottonwood trees and when the air picks up and blows hard they rise up off the branches, float for a second or two, ruffle their feathers, then land again. I listen to the wind, up high, and it is the noise of pheasants and quail, scattering and pouncing into the air when roused. The clouds, this early spring day, are floating in formations. I call them goose clouds.

It was a dry winter and so far, a dry spring. I think of rain as I examine the ground. It is so dry and the land is awkwardly and unwillingly cracking apart into a desperate looking jigsaw puzzle, making homes for bugs, not for roots. When I look down to the ground, I think of earthworms. What sort of tales do they tell? When too much rain falls they emerge for fear of drowning. In a land of dirt and darkness, of water and rivers and springs, the thought of drowning in the soil seems contrary. I wonder, what does the earth look like when five-hundred year floods occur, or when the earth has dried up over geologic time. I wonder, how many earthworms must have surfaced, or perished, at these moments, and of what size. What must be the girth of the world's largest earthworms when they have risen during our earth's largest, deepest floods.

Today, finally, we will get rain. This morning the light is darkening and the sky is becoming the color of dirt. Deep, gritty, black clouds are lowering and preparing to lay down the rain of a thousand lifetimes. The thunder is pounding in the background and lightning is on its way. I think there ought to be an atmospheric division of archeology. Afterall, more of the world is above the ground than below.

It continues to darken, the middle of the day is upon us and the light outside is nearly that of nighttime. Everything seems separated in a storm, everything has to stand alone during harsh weather. The rain is falling now. I look forward to seeing the land drink.

The Weather Movie

The time: A small child in Iowa, alone, alongside an empty lot where my group of friends played football, rode bikes, threw mud balls, heaved apples, kicked field goals, eventually rode skateboards and sometimes waited for the postman to have rubber band fights with.

The camera pans over, slowly: I am on my back, closer to sitting up than laying flat, on top of a large pile of grass, cut branches and leaves. The pile of stuff is emitting a slight odor like apple cider. Eventually this mound of stuff will be deep, dark, crumbly compost dirt.

Laying in this spot years ago, about age eight, I first took a liking to the weather and sounds of nature. I remember laying there in the early morning during autumn when the air was fresh, cold and brisk. The sun was up, but I needed to snuggle into the pile of grass and clutch my sweatshirt to stay really warm. I looked up at the sky, in a rare moment of silence for me back then, and thought about the stillness. I noticed the orange-red tinged sky and then I heard the sound of a mourning dove. This noise sounded so much like the feelings I had inside me, warm and safe, looking out at the wide world with sympathy and curiosity. I thought about how, if I ever made a movie it would start like this: me resting in a pile of warm, cut grass and fall leaves in the early morning. Puffs of steam coming out of my mouth and nose, looking at the sky, with the slow, rich, cooing sound of a mourning dove in the background.

Knot Kansas Zen: Plain Aura

First off, I'm from Kansas and want to make it clear that I want nothing to do with Zen. I live with time, not separate of it. I move into time as clearly as the broad side of a barn shows itself to a southern, summer sky. As obviously as a field of sunflowers looking at the sun. Life is too damned short to go fast, or to step to the side of. I declare Zen out of bounds. I imitate Coyote for bold independence. I am straight as an arrow. I use cattail tufts for my pillow filling, open air as a blanket.

Like I said, the place is Kansas. Where the sea hides silently, layered, beneath years of soil, alongside old, majestic, mountains called the Nehamas, which once rose higher than the Rockies. It is now the land of short and tall grasses, it is a forgotten frontier. There are things here that most people miss when they visit or pass through. In fact, it's possible that even people who live here their whole life miss most of what is being offered. Some locals say they can paint the land, but do so like the land was plastic, as if there weren't gullies and potholes and stuff to trip in. Other people drive through and contemplate if the folks in Kansas are really as mean as what they see of the land. Wonder if we're hazy, pale, humid and full of air. A lady from the cozy cascades of Oregon once asked me if I liked to eat lead shot. I grumbled, "Sure, it's a royal delicacy when mixed with fresh meat."

Recently I listened to a Zen Monk talk about the workings of Zen: how the practice of Zen comes and goes like the wind, how one handles empty space and feels at-one-ment within time. Then he went on to talk about backwards principles and being night during day or some such thing. Keeping in mind that I'm strange myself, what the Zen Monk was saying sounded like a crock of crap to me. Don't get me wrong either, crap is useful, but when it comes out of the mouth though, it's a sorry ass excuse for anything good. Everyone knows where crap really comes from and so what I was hearing made me think that the Zen man was the one thinking backwards. I knew one thing, he'd never made compost before.

As for myself, I have lived off the wilderness in Kansas four-hundred and sixty five years. During this time, I have drawn a detailed diagram of the workings of the brain, mapped out the position of the stars, raced the north wind and witnessed the rise and fall of men countless times. I have sung and heard many songs that left welts on my tongue and sores in my ears.

Recently I figured out something about the structure of wood. This is what I want to tell you about now: how wind and space and time can mix to form a knot.

Around here, time is the most important thing. You must wait for everything. This is what I call see-through time. With see-through time all things are constantly changing. But mostly there is continuity. After all, a change is no more than a step out of the ordinary. For instance, if you were to watch a grove of trees on the horizon, first you would notice the obvious–the stand of trees and that they all seem about the same height. Then if you continued to gaze you would begin to notice changes: wind, birds, falling limbs, or even the chipping of bark. If I am to explain see-through time it is to be expected that it will take some telling and that clarity will arrive only after a lengthy explanation. Bear with me, a thing that's instantaneous is a thing that equals misdirection. Believability is a judgment one makes following patience. In my particular explanation of time, there is a sort of playful lying which will need to be considered as well.

Process is time. People need to remember that things don't *just* happen. Everything occurs slowly, developing ever-so slightly, and so there's a story everywhere and in anything you pick up, or think about. Because I have lived so long, I have witnessed big bluestem grass take root and cover the state. I have watched lightning set the grass ablaze and then watched everything grow back again. I have watched the prairie dry up, wither and slowly return to a lush field of green again. As with anything, the drought and the dust were not the end, but only a development. There is no start, or end to anything.

Simple as it may seem, this continuity is hard to grasp. I often try to explain this delicate fulcrum, this tenuous, ferocious and complex movement of time to others, but instead I get a lot of rolling eyes and stupid comments about my age and state of mental health. There seems a desire, in the mind's of most people, not to believe in difficult or complex ideas. People would rather be shielded from the world and its deep mystic patterns. They simply want to know about simple, quick human creations. Anything natural, or native, or slow, is considered utter madness and insanity.

There are many stories I could tell, for I have seen antelope running inside a gust of wind, rivers emerge from trees, rabbits sprout hooves, thunderstorms living inside earthworms, and lightning hatch from eggs.

I have, of late, been lucky to find the crystal ball of see-through time and am beginning to feel the soul of life. This center of nature is hatching and growing in the heartwood of the Osage Orange tree. Some call this tree Ironwood, others rant that it is the Hammer of Thor. This tree is perhaps one of the most essential strengths of Kansas. Sunshine lives and breathes inside this bright yellow wood. It is a material that cannot be skipped over. It is brilliant, strong and consistent.

Think of it like this: cut red rock in Utah and you will think of blood. This, I understand, is a good sign that you are beginning to understand the land. In Kansas if you cut an Osage Orange tree, you would be wise to try and feel the earth beating, the sun shining, the grass growing deep into the earth. At first try to feel the sun and the shades of prairie grass. When you begin to feel sturdy and safe in the middle of an open, boundlessly arid spot, where the wind is blowing fiercely from the west, recognize you are becoming like the land. You are becoming like a trunk of wisdom, rooted in the heartwood of the Osage Orange. You may even hear the wind blowing, against your ear, whispering, encouragement. If you do hear noises, or if you feel inclined to bow respectfully to the wind, you won't be the first to do so. Just as you won't be the first to find yourself feeling wonderful, turning your head slowly, and looking clearly into the middle of where you are, somewhere close to lost, surrounded and at the apex of the horizon, nestled within the angle of mirage, able to set your consciousness free.

The Osage Orange stands in rows and is much like a horizontal totem pole across the land. And believe me, there are many visions to be seen in a row of Osage Orange trees. Some of us in Kansas collect the green hedge apples that fall off the Osage Orange, squeeze the sap out and boil it with water to make a syrup. When you drink the mixture it helps you listen to the clouds and talk to roots. A group of us did this last summer during the drought which we called "Match Light." It took a lot of drinking, but we finally convinced the sky to let us have a thunderstorm, which drenched us with five full inches of rain.

There are medicine cabinets in the trunks of Osage Orange trees, filled with wisdom, potions and patience. One of my friends, Lester Cluhuk, told me that one time he was cutting branches off the edges of a big old Osage Orange tree when he discovered a hollow spot deep inside the tree's main trunk. As he looked inside the hole, along the edges, he spotted shelves of relics from around the area. There were test tubes of notable high pressures systems, bottles of lake water, noises that blue herons had made in the nearby sycamores. There was a small jar that said catfish on the side, a little pile of meadowlark feathers, a quart of buffalo urine, some lightning bolts, a rainbow, chips of cow dung, all sorts of things really.

Lester also said that he opened up one of the test tubes to see what would happen. He opened up the bottle labeled Cat Fish and as soon as he did, the ground he was standing on turned into a farm pond. The water was deep and cold. Lester said he could feel twenty, maybe thirty big catfish swimming up against his legs. Then, because Lester couldn't swim too well, he started wondering how he could get rid of the water, and get back to the shore. First thing he thought of worked. He put the lid back on the test tube and everything dried up, even his clothes.

It was just recently that the Osage Orange gave me a new thing to talk about. The demonstration and, I suppose, confirmation of see-through time. Mostly it was an invisible feeling, a thing slower than patience could possibly be. I know this only because I have lived so long that I can recognize the clues. You see, it was similar to condensation. I was getting firewood, trimming branches off an Osage Orange tree, when I noticed a hollow tunnel and heard a howling, a deep breathing more like, inside the tree. I peeked down the hollow branch and saw, at the end, a knot of wood being churned. The knot was slick and cool and just like a crystal ball. It was being spun inside a bubble of water and appeared to be blinking, bright, like sunshine inside out. There was a small howling noise rising from the spin of the knot. The knot was being constantly coated and whirled in some kind of highly nutritious, rich water. As the knot turned, it was slowly blinking and with each revolutionary blink, it seemed a decade passed by. I stared, not moving, glued with fascination. The knot was the heart of the prairie.

Vision was not enough. I placed my ear alongside the hollow branch and listened to the flow of time. It was like the empty claw of a fossilized seashell. It was the noise of an ocean, of vast eons of space mixed with animals breathing, eating and playing. I could hear a meadowlark singing, and even recognized the noises of extinct species howling under the earth: the slurp and gulping of a giant sloth crunching leaves, the squawk of a pterodactyl. I noticed the sweet smell of spring honey locust flowers and the silent act of water turning trees into petrified wood.

Water was also dripping out of the open limb, like a cold water spring, forming a small creek down the tree's trunk. Bark trout, if they existed, would have been plentiful alongside the tree. There was a mud puddle forming and my feet had become stuck. I was slowly sinking into the earth. I placed my arm down the tree limb to feel the knot, but quicker than I could respond, I froze up. I looked behind me, out to Kansas and it was wintertime. I was watching time move by quickly, season after season, before my eyes. Three or four blinks and a couple of decades passed by. Time passed smoothly with my arm frozen in the tree limb, my hand placed on the bright yellow knot. I didn't want to move. I was transfixed.

Being transfixed to a tree was not easy. My stomach began to grumble and growl like a dump truck's transaxle. A heavy gear lube felt like it was sloshing deep inside me. I was slowing down until I reached a state of hibernation. I became groggy and hardly aware of my senses. I was halted in place as if a north wind had hit me and laid an anchor at my ankles. If words can capture the feeling, I felt in communion with the molecular structure of a granite boulder.

I maintained my gaze upon the passage of see-through time. I hesitated to look away from the comfort of the knot of wood, especially since it seemed that I was beginning to spread out all over the ground. I could feel my legs beginning to tingle. Where the drops of water had been collecting and where I thought I had felt myself sinking, there were now two giant taproots shooting from my toes, down and out through the ground. I was standing as only a tree could stand, in one place, firmly and with purpose, absorbing, becoming what was nearby. There were earthworms easing past my feet, snuggling against me. At first this didn't seem like that big a deal, the whole process had been so smooth and easy. Then it struck me that soon I wouldn't be able to move, that my skin was turning into bark and that my shoulders were growing hedge apples and small thorns. I cranked together all my remaining human strength and pulled my arms and legs from the prairie soil. I stumbled meekly, pitifully, to the ground.

Being half tree, half person, I began to crave water. I stared into the heart of the yellow spinning Osage Orange knot and was relieved for a moment, but my ankles, and the roots, which had become my feet, were searching for water. I don't know what made me think of it, but I cupped my grainy and woody hands together under the water that was dripping out of the tree limb and gathered water to drink. The results were like a Baptism, setting me free. I felt the roots I had sprouted reeling. I felt them spinning and bouncing back into my body. It felt like warm water flowing over me as the warmth of the Osage Orange's yellow wood entered and found places to live within my body. Finally, I pulled a leaf from my mouth and was able to walk again.

Like I said earlier, I am a bit of a strange one myself, but I do know that I can't possibly make up everything that happens to me. Ever since I swallowed the water coming out of the Osage Orange tree, things have been different. I can talk to insects and turtles. I can grow stalks of corn from the palms of my hands and carry on conversations with the penumbra of the setting sun.

Most of all though, I feel like I am truly a Kansas landscape. I am now wind, space and a broad deception of altitude. I am eighty percent sky and twenty percent land. Within that hundred percent, I hide my water mysteriously. I exist openly. I am not folded inside time. I am simply Kansas. I speak it reverently, with a strong blast of air. I am a plain aura, tied together with shadow and vast amounts of grain. I can stare to the ends of the earth and see through time. Everything is well knotted and secure.

Deep underneath the iron side of stone is the black shadow of coolness. Put your nose down near the ground and smell the scent of sky, of soft, crumbly land. Over time weather, wind, heat, cold, rain and hail, crack hard stone.

Make the land you live beside, walk across, dig in, personal to you. Then get started. Make fire with water. Discover that you can make soil with air, water, grass and apple peels. Alchemy is moving with the elements, wind is moving with the elements, sunsets are moving with the elements, rivers are moving with the elements, humans are moving with the elements, the fire hole in Yellowstone is the movement of the elements. If you want more clues, consult the many religions of the world, the many philosophies, but keep in mind that they are not, any of them, right, for creation is merely the equal of chaos. Creation is never created. Faith is the glue. Perhaps you'd like to take a peek into the sciences, the hard ones, the soft ones, the physics, the chemistry, the psychologies, the botanies. Go ahead, take your time but they are all guess work as well. The work of mystics, fables, fairy tales, these are the best beginnings and what more do any of us have than imagined beginnings? Being correct is not dependent on fact, but on the style and manner in which a story is told. A good story is never complete. A good story is retold and changes and stands the test of time. Stories are the solid harvesting of stable faith.

Imagine a drip of fire. A lamp of air. It may be impossible to become a shaman just because you feel like wanting to be one, but practice to be one anyway. Something, some story, some spirit, will take you in.

Trust your inventiveness and desires. Emerge. Make strong, bold, hard and long lasting efforts to use creation in your life. Invent the world. Along the way, take time to watch water snap stone.

Broom Totem

Zoom in. Zoom yourself in close. Closer.

Look down from where you are at the woman in the field, on the prairie. Such a tiny field, such a tiny house and porch, such a tiny person from high above. Zoom down closer to find her story. Observe.

She stands, planted, grasping the plains gothic tool. Held at her side, prouder than pitchfork, is a broom she has made on her own. The blowing wind is enemy, partner with reap and sow. Sweeping is for her, an introspective movement, intuitive as bristles slide from side to side.

It is maize colored corn stalk, tightly wrapped and bound. Hands held to the handle, at an angle, with the broom stick, the feet dancing, side stepping. Nothing is ever pushed, everything is being smoothly slid. Wood floors cleared and cleared again. Carving herself out she imagines.

Dinner preparation. Cleanly and obvious. Sometimes he, the addition of her husband, brings a flower. What he really knows of her is as narrow as a glint of light in the back of a mouth.

The sun is partner in all this. It brings the life and structure, grows the plants to make the brooms. There is a clearing in the forest, a woodlot, that he goes to and it reminds him of his home. The clearing is all the time much clearer than before.

There is focus pulling in now. The economics of the house: tidy and tight. Slid and rubbed together. Right now it is her on the porch. Broom inside her grip, searching out the distance with her eyes. This searching is both the beginning and the end. For now it shall be the beginning.

The days start early with the sun. She looks as always to the view that now seems to have lasted a thousand life times and thinks back to her childhood, which she also thought would have lasted forever. She has been standing on the porch for at least thirty minutes, not moving an inch out of her mind. When was the moment she left her childhood and entered this present place and time, this adulthood, which doesn't seem to be involved in the future, as much as in an over and over again present. Out in front of her, fanning out in full dimension, the far away equal to the close, spreads the prairie. It is primarily the stepping ground for flatness. Waist high grass and prairie dogs skitter and bark. There is water though, she's discovered, and there are trees as well, but both are folded in between hills and valleys which she really didn't notice in the first few years she lived here. She used to know height through contrast, not through subtleness. Coming from Tennessee, back east, she remained afraid of open space until she found her first pond by accident. Yet, like a trick, this pond dried up in July, revealing no more than bones, dust and weeds.

It was half way into her third year on the prairie before she found the small, spring-fed creek, hidden inside the view off her porch. It wasn't until finding a reliable source of water that she was able to start building a trust with the place she knew as the Great Plains.

During the first years she was simply too scared to move. It was partly the wind that made things seem so alone and wicked. Prairie wind blows at a person stronger and longer and harder than anywhere else since the air gets a jump start by traveling down hill, off mountains, or across broad, uninhabited space from the confines of the arctic circle. There are times when the wind just won't stop and carries a noise that seems to whisper in a person's ear, "mistake, missssstake."

She and her husband are now in love with their lives. They only had dreams of happiness years ago. Things have changed appropriately. They have grown accustomed to what is around them. They have space to live their lives. Funny thing, she thinks, her childhood is beginning to come back to her.

She speaks to her brooms sometimes. She turns one upside down. Stares into the bristles and cries and shakes like a wind storm. She has dust bunnies running through her mouth and around every corner of the farm yard, she believes there is a copperhead waiting to poison her. She envisions the snake up close, in her face. Fangs, vipers, snapping and latching on to her neck. She envisions herself as a small pile of white bones, quiet in the sun. She sits looking into the broom feeling like she is crumbling apart. She brought her grandmother's old rocking chair with her to survive just these kinds of visions though. Such elderly and confident memories flood her from this chair. Fears cease, but not easily. The chair is needed.

She often shook for weeks during the first winter she was on the prairie. She saw her house grow dirty, cold and spooky. A dark blue light would overtake everything, seep through the windows and all the while she just rocked back and forth, shaking.

She would sit, focusing on the bristles of her broom, wondering how the things she once wanted–silence, open views, the feel of grass seeds in her hand–had now become more like ghosts. Every once in a while though, during the early years, her mind would switch. She would tell herself that if only the wind would die down, and the sun heat things up, then everything would be good.

She was prone to forgetting that she'd left her home back east on purpose, in order to come to the prairie. She had envisioned, from far away, that the place would be like living in a dream. During the first days and years it seemed odd that she had wanted to move to the Great Plains since she was a little girl. Her only comforts, it seemed early on, was the view off her front porch and her husband's intense blissfulness at being where he wanted to be, on the prairie, working his own land. More land than a person could walk across on a single, well lit day.

She wondered if she wasn't even more crazy when she was with him though. Afterall, she acted normal when he was around. She used to sing songs to herself when he was gone, about how she was going to be good and that things would be better come morning. She would think about how, when she woke up, she'd try to start all over again. Would look out her window and suddenly awaken to what she had wanted, to live wide and alone on top of the vast and open world.

At times, she vaguely remembered, like cold thick syrup resting between her ears, what she had thought the prairie was going to be like. A place so full of adventure and solitude, well away from people and their opinions. A place where she would have the chance to hide and do whatever she wanted, whenever she wanted. To re-create herself and be isolated. She'd hoped to learn the language of plants and animals. To invent security out of vast, plain, broadness.

She speaks out, thinks out loud. You can hear her as she speaks to a new broom now.

"I mostly see myself as I dance back and forth. I'll stand in front of a mirror and dance with the top of a broom. Brooms are my prairie partners. I named the broom I am dancing with now Wig Head. Dancing away from the mirror I flip Wig Head down and sweep my floors and porch like a new pioneer, like a plow, the way the dirt mounts up so quick along the sides of the bristles. The wind brings dirt inside and I've come to notice that the breeze is noticeable only when it is above twenty miles per hour. Anything less is just constant. I usually clean awhile after he leaves, and then I go out to the porch and look at the view, which is parched dry in the summer, but full of the greenest and biggest leaves that trees have ever grown by the time late spring comes around. Silver maples ruffle white, cottonwoods blow out fluff.

"We are also lucky to have a well hidden creek out by our house. It took me a few years to even find it. We are also lucky to have hills crisscrossing out into the distance to make us feel at home on the range, as the song goes. I look off into the view ten times a day. I have the view under control. I worship it. I planted sunflowers along the creek and there are cattails too. I pretend to carry the creek in my pocket when I have to go to town. This way I can face the folks who make me nervous, all the while, telling myself that I have a safe place out on the prairie.

"I have started to paint pictures of this place, of the view and the prairie. Sometimes I show them to him, and he thinks there are beautiful. He says I should send them to Kansas City, or Omaha, maybe even Chicago. I might. He especially likes the ones that show the beautiful clouds and colors. I like the ones most that focus on the surface of the ground, about how you could fall into a prairiedog hole if you didn't watch where you were walking.

"The secret I have is down by the creek, alongside a narrow fold in the land, beneath the ledge of an open field. I go there when he leaves to work. I walk through the view off our front porch, over a few hills, to the creek. This hidden spot is where I put my sorrow away and where I find hope and courage. I dig graves to bury the hard efforts and bad memories I've had and still get sometimes. Like my first attempt to grow flowers which failed in the prairie heat. I buried the dried out seeds, which never even germinated, in my secret spot, and stuck a broom in the ground, with the words *Seed Dust* on the handle, to mark the spot. But, then, two springs later I noticed the flowers popped up around this broom, proving that things can get better.

"I take a magical approach to digging and burying parts of me and my life: I make up chants, boil my saliva with feathers, write my memories on bark chunks, and stir all sorts of things together like tree sap, deer scat, dried grass tassels and milk weed, making various potions that I experiment with in various ways. Sometimes I even try the potions out on myself. I do all these things on my own and he doesn't know about any of it.

"It is my Broom Graveyard. It is a clean place. It has a permanent planting and growing season. There are broom handles with bristles sprouting all over. If someone were to ever see me in my secret spot it wouldn't matter, because they'd be more in my world than their's and so more lost than me.

"At my secret spot I have cleared a graveyard for my brooms, respecting all that they have cleared for me. Broom handle after broom handle stuck into the earth. I have made so many brooms. Whenever I get a new idea, I stop using the broom I'm holding on to and write the new thought on the handle. I have many memories buried in the fold of the land, many brooms sticking out of the ground at my secret spot. I probably have six hundred brooms, easy, stuck in rows of twenty. I remember them all. Each handle has its own pattern and feel, its own collection of thoughts that come back to me when I touch them."

She stops talking and the wind picks up and you are being taken away from this spot. Away from this woman of the plains, who, up close, is so large, but who is now becoming so small. As you move away and look back, she is barely visible. She blends with the grass, with the water, with the heat of the sun.

All that she is turns to vague and blurry dots. The entire landscape which holds her life, merges together, folds and takes her into the earth. Finally, she is no longer visible. She has been cleared away, swept into the broadness of the prairie earth. She is alone again. You hear her say one more thing as you move away, "Look. Look one more time. I am invisible for I have been changed. I am moving with the elements. I have become the world."

The text of this book has been set in Adobe Garamond. The garamond typeface was designed by Jean Jannon in 1615. Garamond is characterized by little contrast between the thick and thin letter strokes, heavily bracketed serifs and oblique stress. The letterforms are open and round, making the face extremely readable.

This book is printed on 60# Glat, acid free, recycled, supple opaque, natural paper, meeting all library standards.

Book printed by Thomson-Shore, Dexter, Michigan
Book Cover printed by Pinnacle Press, St. Louis, Missouri

Cover and book design assistance provided by
Andy Driscoll and Laura Waldo

Established in 1994, the Ice Cube Press is a small, independent press dedicated to publishing writing on nature, the environment and regionalism.

Also Available:
River Tips and Tree Trunks: Notes and Reflections on Water and Wood.

"As rooted as trees" – *Small Press Review*

"Kicks, Bristles, Barks and Radiates"
– *The Lawrence Daily-Journal World*

"An uplifting spirit!" – Marion Times

To order River Tips and Tree Trunks,
or additional copies of
Moving With The Elements
send $14.95/book (+1.50 handling) to:

The Ice Cube Press & Letterpress
205 North Front Street
North Liberty, Iowa 52317-9302
Or order with your credit card from our web site – http://soli.inav.net/~icecube

Your patronage of this and all small, independent presses is appreciated.